ISBN 0-8373-5313-0

13 COLLEGE LEVEL EXAMINATION SERIES/CLEP

New **RUDMAN'S QUESTIONS AND ANSWERS ON THE...**

CLEP

College-Level Examination Program Subject Test In...

GENERAL CHEMISTRY

Test Preparation Study Guide

Questions and Answers

NLC

NATIONAL LEARNING CORPORATION

National Learning Corporation
212 Michael Drive, Syosset, New York 11791
(516) 921-8888
(800) 645-6337
Fax (516) 921-8743
www.passbooks.com

PRINTED IN THE UNITED STATES OF AMERICA

PASSBOOK®

NOTICE

NATIONAL LEARNING CORPORATION
212 Michael Drive
Syosset, New York 11791

Inquiries re licensing agreements should be addressed to:
The President
National Learning Corporation
212 Michael Drive
Syosset, New York 11791

PASSBOOK SERIES®

THE *PASSBOOK SERIES®* has been created to prepare applicants and candidates for the ultimate academic battlefield—the examination room.

At some time in our lives, each and every one of us may be required to take an examination—for validation, matriculation, admission, qualification, registration, certification, or licensure.

Based on the assumption that every applicant or candidate has met the basic formal educational standards, has taken the required number of courses, and read the necessary texts, the *PASSBOOK SERIES®* furnishes the one special preparation which may assure passing with confidence, instead of failing with insecurity. Examination questions—together with answers—are furnished as the basic vehicle for study so that the mysteries of the examination and its compounding difficulties may be eliminated or diminished by a sure method.

This book is meant to help you pass your examination provided that you qualify and are serious in your objective.

The entire field is reviewed through the huge store of content information which is succinctly presented through a provocative and challenging approach—the question-and-answer method.

A climate of success is established by furnishing the correct answers at the end of each test.

You soon learn to recognize types of questions, forms of questions, and patterns of questioning. You may even begin to anticipate expected outcomes.

You perceive that many questions are repeated or adapted so that you gain acute insights, which may enable you to score many sure points.

You learn how to confront new questions, or types of questions, and to attack them confidently and work out the correct answers.

You note objectives and emphases, and recognize pitfalls and dangers, so that you may make positive educational adjustments.

Moreover, you are kept fully informed in relation to new concepts, methods, practices, and directions in the field.

You discover that you are actually taking the examination all the time: you are preparing for the examination by "taking" an examination, not by reading extraneous and/or supererogatory textbooks.

In short, this PASSBOOK®, used directedly, should be an important factor in helping you to pass your test.

College-Level Examination Program (CLEP)

1. WHAT IS CLEP?

CLEP stands for the College-Level Examination Program, sponsored by the College Entrance Examination Board. It is a national program of credit-by-examination that offers you the opportunity to obtain recognition for college-level achievement. No matter when, where, or how you have learned -- by means of formal or informal study -- you can take CLEP tests. If the results are acceptable to your college, you can receive credit.

You may not realize it, but you probably know more than your academic record reveals. Each day you, like most people, have an opportunity to learn. In private industry and business, as well as at all levels of government, learning opportunities continually occur. If you read widely or intensively in a particular field, think about what you read, discuss it with your family and friends, you are learning. Or you may be learning on a more formal basis by taking a correspond nce course, a television or radio course, a course recorded on tape or cassettes, a course assembled into programmed tests, or a course taught in your community adult school or high school.

No matter how, where, or when you gained your knowledge, you may have the opportunity to receive academic credit for your achievement that can be counted toward an undergraduate degree. The College-Level Examination Program (CLEP) enables colleges to evaluate your achievement and give you credit. A wide range of college-level examinations are offered by CLEP to anyone who wishes to take them. Scores on the tests are reported to you and, if you wish, to a college, employer, or individual.

2. WHAT ARE THE PURPOSES OF THE COLLEGE-LEVEL EXAMINATION PROGRAM?

The basic purpose of the College-Level Examination Program is to enable individuals who have acquired their education in non-traditional ways to demonstrate their academic achievement. It is also intended for use by those in higher education, business, industry, government, and other fields who need a reliable method of assessing a person's educational level.

Recognizing that the real issue is not *how* a person has acquired his education but *what* education he has, the College-Level Examination Program has been designed to serve a variety of purposes. The basic purpose, as listed above, is to enable those who have reached the college level of education in non-traditional ways to assess the level of their achievement and to use the test results in seeking college credit or placement.

In addition, scores on the tests can be used to validate educational experience obtained at a nonaccredited institution or through noncredit college courses.

Some colleges and universities may use the tests to measure the level of educational achievement of their students, and for various institutional research purposes.

Other colleges and universities may wish to use the tests in the admission, placement, and guidance of students who wish to transfer from one institution to another.

Businesses, industries, governmental agencies, and professional groups now accept the results of these tests as a basis for advancement, eligibility for further training, or professional or semi-professional certification.

Many people are interested in the examination simply to assess their own educational progress and attainment.

The college, university, business, industry, or government agency that adopts the tests in the College-Level Examination Program makes its own decision about how it will use and interpret the test scores. The College Board will provide the tests, score them, and report the results either to the individuals who took the tests or the college or agency that administered them. It does NOT, and cannot, award college credit, certify college equivalency, or make recommendations regarding the standards these institutions should establish for the use of the test results.

Therefore, if you are taking the tests to secure credit from an institution, you should FIRST ascertain whether the college or agency involved will accept the scores. Each institution determines which CLEP tests it will accept for credit and the amount of credit it will award. If you want to take tests for college credit, first call, write, or visit the college you wish to attend to inquire about its policy on CLEP scores, as well as its other admission requirements.

The services of the program are also available to people who have been requested to take the tests by an employer, a professional licensing agency, a certifying agency, or by other groups that recognize college equivalency on the basis of satisfactory CLEP scores. You may, of course, take the tests SOLELY for your own information. If you do, your scores will be reported only to you.

While neither CLEP nor the College Board can evaluate previous credentials or award college credit, you will receive, with your scores, basic information to help you interpret your performance on the tests you have taken.

3. WHAT ARE THE COLLEGE-LEVEL EXAMINATIONS?

In order to meet different kinds of curricular organization and testing needs at colleges and universities, the College-Level Examination Program offers two kinds of examinations at its test centers -- the *General Examinations* and the *Subject Examinations*.

4. WHAT ARE THE GENERAL EXAMINATIONS?

The General Examinations are objective tests, of the multiple-choice type, that measure achievement in five (5) basic areas of the liberal arts:

English Composition
Humanities
Mathematics
Natural Sciences
Social Sciences - History

The General Examinations are intended to provide a comprehensive measure of undergraduate achievement in the five (5) basic areas of the liberal arts. They are not intended to measure advanced training in any specific discipline, but rather to assess a student's knowledge of fundamental facts and concepts, his ability to perceive relationships, and his understanding of the basic principles of the subject. The content of the General Examination is similar to the content of those subjects ordinarily included in the program of study required of most general education students in the first two years of college.

Each of the General Examinations has a time limit of ninety (90) minutes. The total battery of five (5) tests requires five (5) hours of testing time plus additional time for instructions and distributions of materials. When the entire battery is given, the tests are usually administered in two (2) separately timed sessions. No more than four (4) ninety (90) minute tests may be taken in one day.

The General Examinations are made up of a number of questions or items, each of which has five (5) answer choices labeled (A), (B), (C), (D), and (E). You will be asked to select the alternative that best answers each question and record your answer by blackening the appropriate space on the answer sheet. Except for the English composition version with essay, each General Examination consists entirely of multiple-choice questions.

You are not expected to finish each test. Many who score well on the tests do not finish them. It is important, however, that you budget your time and pace yourself throughout the time allotted in order to answer as many questions as possible. Because there is a time limit for the test and because of the correction for guessing (explained below), you would probably be wise to skip the more difficult or perplexing questions and return to them if there is time later.

Each time you mark an answer, you should be certain that the number on the answer sheet is the same as the number of the questions.

Although the multiple-choice items in these tests are carefully designed so as not to be tricky, misleading, or ambiguous, on the other hand, they are not all direct questions of factual information. They attempt, in their way, to elicit a response that indicates your knowledge or lack of knowledge of the material in question or your ability or inability to use or interpret a fact or idea. Thus, you should concentrate on answering the questions as they appear to be without attempting to out-guess the testmakers.

Many individuals wonder whether or not to guess the answers to questions about which they are not certain. In these tests, a percentage of the wrong answers will be subtracted from the number of right answers as a correction for haphazard guessing. Omitted items are not counted as either right or wrong. It is highly improbable, therefore, that mere guessing will improve your score significantly; it may even lower it, and it does take time. If, however, you are not sure of the correct answer, but have some knowledge of the question and are able to eliminate one or more of the answer choices as wrong, your chance of getting the right answer is improved, and it may be to your advantage to answer such a question.

Scores on the General Examinations are derived by counting the number of correct and incorrect responses and subtracting a percentage of the wrong answers to allow for guessing. These scores are then converted to a scale that, for the basic scores, ranges from a low of 200 to a high of 800, with a mean (or average) score of 500. For subscores, the scale is 20 to 80, with a mean score of 50.

You do not have to answer all, or even most, of the questions correctly to earn acceptable scores on these tests.

Scores on the General Examinations have no fixed value; there are no passing or failing score levels established by the College Board. Your score reflects your standing in relation to the scores of others who have taken the tests. Some colleges, universities, and other agencies, however, may require particular score levels for awarding credit or for advancement. Three or six hours of credit are usually awarded for satisfactory scores on each General Examination.

5. WHAT ARE THE SUBJECT EXAMINATIONS?

Each Subject Examination covers material taught in an undergraduate course with a similar title at most colleges and universities. Each is ninety (90) minutes long and is composed of multiple choice questions to be answered in two separately timed sections. Most include a separate 90-minute essay section that should be taken only if required by the institution receiving your scores. Please note that the Program does not provide a scoring service for the essay question; these are graded by the designated recipient. No more than four (4) Subject Examinations without essay, or two with essay, may be taken in one day. The Subject Examinations that have been available are taken from the list on the following page:

- American Government
- American History I, II
- American Literature
- Analysis & Interpretation of Literature
- Biology (General)
- College Algebra
- College Algebra-Trigonometry
- College Composition
- Computers & Data Processing
- Educational Psychology
- Elementary Computer Programming
- English Literature
- General Chemistry
- General Psychology
- Geology
- History of American Education
- Human Growth & Development
- Introduction to Business Management
- Introductory Accounting
- Introductory Business Law
- Introductory Calculus (with Elementary Functions)
- Introductory Economics
- Introductory Marketing
- Introductory Sociology
- Money & Banking
- Statistics
- Tests & Measurements
- Trigonometry
- Western Civilization I, II
- Fundamentals of Nursing
- Freshman English
- Clinical Chemistry
- Hematology
- Immunohematology & Blood Banking
- Microbiology
- Afro-American History
- Medical-Surgical Nursing
- Anatomy, Physiology, and Microbiology
- Behavioral Sciences for Nurses
- Introductory Microeconomics
- Introductory Macroeconomics
- Introductory Micro- and Macro-economics
- Calculus with Analytical Geometry
- French (College Levels I & II)
- German (College Levels I & II)
- Spanish (College Levels I & II)
- Dental Auxiliary Education Examination in
  - Dental Materials
  - Head, Neck & Oral Anatomy
  - Oral Radiography
  - Tooth Morphology & Function
- Information Systems and Computer Applications

The Subject Examinations are essentially end-of-course tests developed for widely taught undergraduate courses. The examinations measure understanding of basic facts and concepts, as well as the ability to apply such understandings to the solution of problems and the interpretation of materials. Questions that require of a student only rote recall are avoided.

The Council on College-Level Examinations has endorsed a model policy statement for use for the Subject Examinations.

On the theory that credit by examination should be granted at least at the C level -- the same level at which credit is normally granted to transfer students -- the Council recommends that credit based on CLEP Subject Examinations be granted to individuals who earn scores at or above the mean score for C students on the CLEP national norms. This is the mean score earned by regularly enrolled college students who participated in the national norms sample and received a grade of C in their college course in the subject concerned.

6. HOW DO THE EXAMINATIONS DIFFER?

The Subject Examinations differ from the General Examinations in that the former are more closely tied to course content and are intended to cover material that is typical of college courses in these subjects. These examinations require a higher degree of specialized knowledge and training.

The General Examinations are comprehensive in nature and are not based on a particular curriculum or course of study. They do not attempt to measure the outcomes of specialized courses that students might pursue when majoring in a particular field. Rather, they sample widely the content of the major disciplines with which each is concerned, stressing understanding, not merely retention, of facts, the ability to perceive relationships, and the grasp of basic principles and concepts.

7. WHAT ARE THE FEES?

The fee at the test centers for each examination is $40.00.

8. WHEN ARE THE TESTS GIVEN?

CLEP tests are normally administered during the middle of each month, and candidates should register for them three weeks before the test date.

9. WHERE ARE THE TESTS GIVEN?

More than 1,000 test centers are located on college and university campuses throughout the country, and additional centers are being established to meet increased needs. Any accredited collegiate institution with an explicit and publicly available policy of credit by examination can become a CLEP test center. To obtain a list of these centers, free of charge, write to: CLEP, CN6601, Princeton, NJ 08541-6601.

10. HOW DO I REGISTER FOR THE COLLEGE-LEVEL EXAMINATION PROGRAM?

Send $40.00 per test to the test center you have selected. Test fees must be paid by check or money order payable to College-Level Examination Program.

11. MAY I REPEAT THE COLLEGE-LEVEL EXAMINATIONS?

You may repeat any examination providing at least six months have passed since you were last administered this test. If you repeat a test within a period of time less than six months, your scores will be cancelled and your fees forfeited. To repeat a test, check the appropriate space on the registration form.

12. WHEN MAY I EXPECT MY SCORE REPORTS?

Your scores, and a booklet that explains them, will be sent only to you and to the recipient you specify within one month after the test date. Copies of essays are sent to and graded by the college. Scores and essays are kept on file for twenty years. During that period, another score report and essay copy can be sent to anyone you designate by filling out the request form included with your score report, or by writing to CLEP Transcript Service, CN 6600, Princeton, New Jersey 08541-6601. Give your name, social security number, the date and place you took CLEP tests, and the name and address to which the transcript is to be sent. Enclose $10.00 for each transcript requested. (A transcript is a report to ONE recipient of the scores earned at a single administration.)

13. WHAT ARE EXTERNAL DEGREES?

Do you want a college degree or diploma without setting foot in a classroom, without having to leave, or take time off from, your job? Yes, it can be done.

Today the word is *EXTERNAL DEGREE*.

For the first time in educational history, qualified individuals can earn college credit or other educational advantages without actual classroom attendance.

And these credits may now be applied directly toward degrees awarded under various external degree programs developed by several state education departments and over 900 cooperating colleges and universities who have agreed to award credit to candidates submitting CLEP scores.

One of these is the New York State Board of Regents, which is providing an alternate approach to a college degree for those who choose to learn on their own. The Regents are now offering the following External Degrees: Associate in Arts, Bachelor of Science in Business Administration, and Associate in Applied Science in Nursing.

The Regents External Degree is a truly *external* degree; it is awarded by a university which evaluates a student it has not directly taught. The University of the State of New York has no campus, resident faculty, or students in the traditional sense. It publishes its requirements and awards a degree to anyone who can meet them. There are no requirements of admission, residence, or age, and the methods of preparation are not prescribed. Degree requirements can be satisfied in several ways. No classroom attendance is required, and no instruction is provided.

Credits may be earned by passing New York State's College Proficiency Examinations or the College Board's College-Level Examinations.

Credit may be earned toward the associate in arts program and the general education component of the business and nursing degree programs by successfully completing any or all of the CLEP General Examinations. The examination title, minimum acceptable scores, and the number of semester hours allowed for each of the examinations are listed below.

COLLEGE-LEVEL EXAMINATION PROGRAM
GENERAL EXAMINATIONS

Examination Title	Minimum Acceptable Total Score	Semester Hours Allowed
English	427	3
Humanities	422	6
Mathematics	424	6
Natural Sciences	422	6
Social Sciences-History	422	6

As for the CLEP Subject Examinations, a candidate must achieve a minimum standard score of 45 on one of these to receive credit toward a Regents External Degree.

Candidates for any of the Regents External Degrees should apply to: Regents College Degrees
Cultural Education Center
Empire State Plaza
Albany, New York 12230

14. HOW SHOULD I PREPARE FOR THE COLLEGE-LEVEL EXAMINATIONS?

This book has been specifically designed to prepare candidates for these examinations. It will help you to consider, study, and review important content, principles, practices, procedures, problems, and technics in the form of varied and concrete applications.

15. QUESTIONS AND ANSWERS APPEARING IN THIS PUBLICATION

The College-Level Examinations are offered by the College Entrance Examination Board. Since copies of past examinations have not been made available, we have used equivalent materials, including questions and answers, which are highly recommended by us as an appropriate means of preparing for these examinations.

If you need additional information about CLEP Examinations, write: CLEP
CN 6600
Princeton, New Jersey 08541-6600.

THE COLLEGE-LEVEL EXAMINATION PROGRAM

How The Program Works

CLEP examinations are administered at many colleges and universities across the country, and most institutions award college credit to those who do well on them. The examinations provide people who have acquired knowledge outside the usual educational settings the opportunity to show that they have learned college-level material without taking certain college courses.

The CLEP examinations cover material that is taught in introductory-level courses at many colleges and universities. Faculties at individual colleges review the tests to ensure that they cover the important material taught in their courses. Colleges differ in the examinations they accept; some colleges accept only two or three of the examinations while others accept nearly all of them.

Although CLEP is sponsored by the College Board and the examinations are scored by Educational Testing Service (ETS), neither of these organizations can award college credit. Only accredited colleges may grant credit toward a degree. When you take a CLEP examination, you may request that a copy of your score report be sent to the college you are attending or plan to attend. After evaluating your scores, the college will decide whether or not to award you credit for a certain course or courses, or to exempt you from them. If the college gives you credit, it will record the number of credits on your permanent record, thereby indicating that you have completed work equivalent to a course in that subject. If the college decides to grant exemption without giving you credit for a course, you will be permitted to omit a course that would normally be required of you and to take a course of your choice instead.

The General Examinations

Each of the five General Examinations covers material taught in courses that most students take as requirements during their first two years of college. Most colleges require all their students to take three or six semester hours (or the equivalent) in each of the five areas. They may grant credit in the area in which a satisfactory test score is earned rather than for a specific course. General Examinations are available in the following areas: English Composition, Humanities, Mathematics, Natural Sciences, and Social Sciences and History.

The Subject Examinations

Each Subject Examination includes material usually covered in an undergraduate course with a similar title. On the basis of a satisfactory test score, each institution grants credit for an examination equal to the amount it gives to students who successfully complete that course. Over thirty Subject Examinations in several curriculum areas are offered. The Subject Examinations are listed on the contents page.

What the Examinations Are Like

The examinations consist of multiple-choice questions to be answered within a 90-minute time limit. There are two versions of the General Examination in English Composition: One is made up entirely of multiple-choice questions; the other has a 45-minute multiple-choice section and a 45-minute essay section. Additional information about the length, timing, and content of each CLEP examination is given in the examination guide.

Optional 90-minute free-response or essay sections are available for many of the Subject Examinations. These sections are required by some colleges in addition to the 90-minute multiple-choice test. The optional sections are not graded at Educational Testing Service (ETS); rather, they are sent to the college designated by the candidate to be graded by local faculty. Before registering for a CLEP examination, you should find out whether the college that is to receive your score requires the optional free-response section.

Where To Take the Examinations

CLEP examinations are administered throughout the year at the test centers of approximately 1,200 colleges and universities. On pages A1-12, you will find a list of institutions that award credit for satisfactory scores on CLEP examinations. Some colleges administer CLEP examinations to their own students only. Other institutions on the list administer the tests to anyone who registers to take them. If your college does not administer the tests, contact the test center at an asterisked college in your area for information about its testing schedule.

Once you have been tested, your answer sheet will be sent to ETS for scoring. Your score report will be mailed a few weeks later, both to you and to the college or other recipient you designate. CLEP scores are kept on file at ETS for 20 years; and during this period, for a small fee, you may have your transcript sent to another college or to anyone else you specify. (Your scores will never be sent to anyone without your approval.)

Active military personnel may take CLEP examinations without charge through the Defense Activity for Non-Traditional Education Support (DANTES) program. Information about CLEP may be obtained from the Test Control Officers at military installations.

APPROACHING A COLLEGE ABOUT CLEP

The following sections provide a step-by-step approach to learning about the CLEP policy at a particular college or university. The person or office that can best assist students desiring CLEP credit may have a different title at each institution, but the following guidelines will lead you to information about CLEP at any institution.

Adults returning to college often benefit from special assistance when they approach a college. Opportunities for adults to return to formal learning in the classroom are now widespread, and colleges and universities have worked hard to make this a smooth process for older students. Many colleges have established special service offices that are staffed with trained professionals who understand the kinds of problems facing adults returning to college. If you think you might benefit from such assistance, be sure to find out whether these services are available at your college.

How to Apply for College Credit

STEP 1. Obtain the General Information Catalog and a copy of the CLEP policy from the colleges you are considering. If you have not yet applied for admission, ask for an admissions application form too.

Information about admissions and CLEP policies can be obtained by writing or visiting college admissions offices. Tell the admissions officer that you are a prospective student and that you are interested in applying for admission and CLEP credit. Ask for a copy of the publication in which the college's complete CLEP policy is explained. Also get the name and the telephone number of the person to contact in case you have further questions about CLEP.

At this step, you may wish to obtain information from external degree colleges. Many adults find that such colleges suit their needs exceptionally well.

STEP 2. If you have not already been admitted to the college you are considering, look at its admission requirements for undergraduate students to see if you can qualify.

This is an important step because if you can't get into college, you can't get college credit for CLEP. Nearly all colleges require students to be admitted and to enroll in one or more courses before granting the students CLEP credit.

Virtually all public community colleges and a number of four-year state colleges have *open admission* policies for in-state students. This usually means that they admit anyone who has graduated from high school or has earned a high school equivalency diploma.

If you think you do not meet the admission requirements, contact the admissions office for an interview with a counselor. Colleges do sometimes make exceptions, particularly for adult applicants. State why you want the interview and ask what documents you should bring with you or send in advance. (These materials may include a high school transcript, transcript of previous college work, completed application for admission, etc.) Make an extra effort to have all the information requested in time for the interview.

During the interview, relax and be yourself. Be prepared to state honestly why you think you are ready and able to do college work. If you have already taken CLEP examinations and scored high enough to earn credit, you have shown that you are able to do college work. Mention this achievement to the admissions counselor because it may increase your chances of being accepted. If you have not taken a CLEP examination, you can still improve your chances of being accepted by describing how your job training or independent study has helped prepare you for college-level work. Tell the counselor what you have learned from your work and personal experiences.

STEP 3. Evaluate the college's CLEP policy.

Typically, a college lists all its academic policies, including CLEP policies, in its general catalog. You will probably find the CLEP policy statement under a heading such as *Credit-by-Examination*, *Advanced Standing*, *Advanced Placement*, or *External Degree Program*. These sections can usually be found in the front of the catalog.

Many colleges publish their credit-by-examination policies in a separate brochure, which is distributed through the campus testing office, counseling center, admissions office, or registrar's office. If you find a very general policy statement in the college catalog, seek clarification from one of these offices.

Review the material in the section of this guide entitled *Questions to Ask About a College's CLEP Policy*. Use these guidelines to evaluate the college's CLEP policy. If you have not yet taken a CLEP examination, this evaluation will help you decide which examinations to take and whether or not to take the free-response or essay portion. Because individual colleges have different CLEP policies, a review of several policies may help you decide which college to attend.

STEP 4. If you have not yet applied for admission, do so early.

Most colleges expect you to apply for admission several months before you enroll, and it is essential that you meet the published application deadlines. It takes time to process your application for admission; and if you have yet to take a CLEP examination, it will be some time before the college receives and reviews your score report. You will probably want to take some, if not all, of the CLEP examinations you are interested in before you enroll so you know which courses you need not register for. In fact, some colleges require that all CLEP scores be submitted before a student registers.

Complete all forms and include all documents requested with your application(s) for admission. Normally, an admissions decision cannot be reached until all documents have been submitted and evaluated. Unless told to do so, do not send your CLEP scores until you have been officially admitted.

STEP 5. Arrange to take CLEP examination(s) or to submit your CLEP score(s).

You may want to wait to take your CLEP examinations until you know definitely which college you will be attending. Then you can make sure you are taking tests your college will accept for credit. You will also be able to request that your scores be sent to the college, free of charge, when you take the tests.

If you have already taken CLEP examinations, but did not have a copy of your score report sent to your college, you may request the College Board to send an official transcript at any time for a small fee. Use the Transcript Request Form that was sent to you with your score report. If you do not have the form, you may write to CLEP Transcript Service, The College Board, P.O. Box 6600, Princeton, New Jersey 08541-6600.

Your CLEP scores will be evaluated, probably by someone in the admissions office, and sent to the registrar's office to be posted on your permanent record once you are enrolled. Procedures vary from college to college, but the process usually begins in the admissions office.

STEP 6. Ask to receive a written notice of the credit you receive for your CLEP score(s).

A written notice may save you problems later, when you submit your degree plan or file for graduation. In the event that there is a question about whether or not you earned CLEP credit, you will have an official record of what credit was awarded. You may also need this verification of course credit if you go for academic counseling before the credit is posted on your permanent record.

STEP 7. Before you register for courses, seek academic counseling.

A discussion with your academic advisor can prevent you from taking unnecessary courses and can tell you specifically what your CLEP credit will mean to you. This step may be accomplished at the time you enroll. Most colleges have orientation sessions for new students prior to each enrollment period. During orientation, students are usually assigned an academic advisor who then gives them individual help in developing long-range plans and a course schedule for the next semester. In conjunction with this counseling, you may be asked to take some additional tests so that you can be placed at the proper course level.

<u>External Degree Programs</u>

If you have acquired a considerable amount of college-level knowledge through job experience, reading, or noncredit courses, if you have accumulated college credits at a variety of colleges over a period of years, or if you prefer studying on your own rather than in a classroom setting, you may want to investigate the possibility of enrolling in an external degree program. Connecticut, New Jersey, and New York offer external degree programs that allow you to earn a degree by passing examinations (including CLEP), transferring credit from other colleges, and demonstrating in other ways that you have satisfied the educational requirements. No classroom attendance is required, and the programs are open to out-of-state candidates as well as residents. Thomas A. Edison State College in New Jersey and Charter Oaks College in Connecticut are fully accredited independent state colleges; the New York program is part of the state university system and is also fully accredited. If you are interested in exploring an external degree, you can write for more information to:

Charter Oak College
The Exchange, Suite 171
270 Farmington Avenue
Farmington, CT 06032-1909

Regents External Degree Program
Cultural Education Center
Empire State Plaza
Albany, New York 12230

Thomas A. Edison State College
101 West State Street
Trenton, New Jersey 08608

Many other colleges also have external degree or weekend programs. While they often require that a number of courses be taken on campus, the external degree programs tend to be more flexible in transferring credit, granting credit-by-examination, and allowing independent study than other traditional programs. When applying to a college, you may wish to ask whether it has an external degree or weekend program.

Questions to Ask About a College's CLEP Policy

Before taking CLEP examinations for the purpose of earning college credit, try to find the answers to these questions:

1. Which CLEP examinations are accepted by this college?

A college may accept some CLEP examinations for credit and not others - possibly not the one you are considering. The English faculty may decide to grant college English credit based on the CLEP College Composition examination, but not on the Freshman English examination. Or, the mathematics faculty may decide to grant credit based on the General Examination in Mathematics to non-mathematics majors only, requiring majors to take a Subject Examination in algebra, trigonometry, or calculus to earn credit. For these reasons, it is important that you know the specific CLEP tests for which you can receive credit.

2. Does the college require the optional free-response (essay) section as well as the objective portion of the CLEP examination you are considering? Or will you be required to pass a departmental test such as an essay, laboratory, or oral examination in addition to the CLEP multiple-choice examination?

Knowing the answers to these questions ahead of time will permit you to schedule the optional free-response or departmental examination when you register to take your CLEP examination.

3. Is credit granted for specific courses? If so, which ones?

You are likely to find that credit will be granted for specific courses and the course titles will be designated in the college's CLEP policy. It is not necessary, however, that credit be granted for a specific course in order for you to benefit from your CLEP credit. For instance, at many liberal arts colleges, all students must take certain types of courses; these courses may be labeled the *core curriculum*, *general education requirements*, *distribution requirements*, or *liberal arts requirements*. The requirements are often expressed in terms of credit hours. For example, all students may be required to take at least six hours of humanities, six hours of English, three hours of mathematics, six hours of natural science, and six hours of social science, with no particular courses in these disciplines specified. In these instances, CLEP credit may be given as *6 hrs. English credit* or *3 hrs. Math credit* without specifying for which English or mathematics courses credit has been awarded. In order to avoid possible disappointment, you should know before taking a CLEP examination what type of credit you can receive and whether you will only be exempted from a required course but receive no credit.

4. How much credit is granted for each examination you are considering, and does the college place a limit on the total amount of CLEP credit you can earn toward your degree?

Not all colleges that grant CLEP credit award the same amount for individual tests. Furthermore, some colleges place a limit on the total amount of credit you can earn through CLEP or other examinations. Other colleges may grant you exemption but no credit toward your degree. Knowing several colleges' policies concerning these issues may help you decide which college you will attend. If you think you are capable of passing a number of CLEP examinations, you may want to attend a college that will allow you to earn credit for all or most of them. For example, the state external degree programs grant credit for most CLEP examinations (and other tests as well).

5. What is the required score for earning CLEP credit for each test you are considering?

Most colleges publish the required scores or percentile ranks for earning CLEP credit in their general catalog or in a brochure. The required score may vary from test to test, so find out the required score for each test you are considering.

6. What is the college's policy regarding prior course work in the subject in which you are considering taking a CLEP test?

Some colleges will not grant credit for a CLEP test if the student has already attempted a college-level course closely aligned with that test. For example, if you successfully completed English 101 or a comparable course on another campus, you will probably not be permitted to receive CLEP credit in that subject, too. Some colleges will not permit you to earn CLEP credit for a course that you failed.

7. Does the college make additional stipulations before credit will be granted?

It is common practice for colleges to award CLEP credit only to their enrolled students. There are other stipulations, however, that vary from college to college. For example, does the college require you to formally apply for or accept CLEP credit by completing and signing a form? Or does the college require you to *validate* your CLEP score by successfully completing a more advanced course in the subject? Answers to these and other questions will help to smooth the process of earning college credit through CLEP.

The above questions and the discussions that follow them indicate some of the ways in which colleges' CLEP policies can vary. Find out as much as possible about the CLEP policies at the colleges you are interested in so you can choose a college with a policy that is compatible with your educational goals. Once you have selected the college you will attend, you can find out which CLEP examinations your college recognizes and the requirements for earning CLEP credit.

DECIDING WHICH EXAMINATIONS TO TAKE

If You're Taking the Examinations for College Credit or Career Advancement

Most people who take CLEP examinations do so in order to earn credit for college courses. Others take the examinations in order to qualify for job promotions or for professional certification or licensing. It is vital to most candidates who are taking the tests for any of these reasons that they be well prepared for the tests they are taking so that they can advance as rapidly as possible toward their educational or career goals.

It is usually advisable that those who have limited knowledge in the subjects covered by the tests they are considering enroll in the college courses in which that material is taught. Those who are uncertain about whether or not they know enough about a subject to do well on a particular CLEP test will find the following guidelines helpful.

There is no way to predict if you will pass a particular CLEP examination, but answers to the questions under the seven headings below should give you an indication of whether or not you are likely to succeed.

1. Test Descriptions

Read the description of the test provided. Are you familiar with most of the topics and terminology in the outline?

2. Textbooks

Examine the suggested textbooks and other resource materials following the test descriptions in this guide. Have you recently read one or more of these books, or have you read similar college-level books on this subject? If you have not, read through one or more of the textbooks listed, or through the textbook used for this course at your college. Are you familiar with most of the topics and terminology in the book?

3. Sample Questions

The sample questions provided are intended to be typical of the content and difficulty of the questions on the test. Although they are not an exact miniature of the test, the proportion of the sample questions you can answer correctly should be a rough estimate of the proportion of questions you will be able to answer correctly on the test.

Answer as many of the sample questions for this test as you can. Check your answers against the correct answers. Did you answer more than half the questions correctly?

Because of variations in course content at different institutions, and because questions on CLEP tests vary from easy to difficult - with most being of moderate difficulty - the average student who passes a course in a subject can usually answer correctly about half the questions on the corresponding CLEP examination. Most colleges set their passing scores near this level, but some set them higher. If your college has set its required score above the level required by most colleges, you may need to answer a larger proportion of questions on the test correctly.

4. Previous Study

Have you taken noncredit courses in this subject offered by an adult school or a private school, through correspondence, or in connection with your job? Did you do exceptionally well in this subject in high school, or did you take an honors course in this subject?

5. Experience

Have you learned or used the knowledge or skills included in this test in your job or life experience? For example, if you lived in a Spanish-speaking country and spoke the language for a year or more, you might consider taking the College Spanish examination. Or, if you have worked at a job in which you used accounting and finance skills, Introductory Accounting would be a likely test for you to take. Or, if you have read a considerable amount of literature and attended many art exhibits, concerts, and plays, you might expect to do well on the General Examination in Humanities.

6. Other Examinations

Have you done well on other standardized tests in subjects related to the one you want to take? For example, did you score well above average on a portion of a college entrance examination covering similar skills, or did you obtain an exceptionally high score on a high school equivalency test or a licensing examination in this subject? Although such tests do not cover exactly the same material as the CLEP examinations and may be easier, persons who do well on these tests often do well on CLEP examinations, too.

7. Advice

Has a college counselor, professor, or some other professional person familiar with your ability advised you to take a CLEP examination?

If your answer was yes to questions under several of the above headings, you probably have a good chance of passing the CLEP examination you are considering. It is unlikely that you would have acquired sufficient background from experience alone. Learning gained through reading and study is essential, and you will probably find some additional study helpful before taking a CLEP examination.

If You're Taking the Examinations to Prepare for College

Many people entering college, particularly adults returning to college after several years away from formal education, are uncertain about their ability to compete with other college students. They wonder whether they have sufficient background for college study, and those who have been away from formal study for some time wonder whether they have forgotten how to study, how to take tests, and how to write papers. Such people may wish to improve their test-taking and study skills prior to enrolling in courses.

One way to assess your ability to perform at the college level and to improve your test-taking and study skills at the same time is to prepare for and take one or more CLEP examinations. You need not be enrolled in a college to take a CLEP examination, and you may have your scores sent only to yourself and later request that a transcript be sent to a college if you then decide to apply for credit. By reviewing the test descriptions and sample questions, you may find one or several subject areas in which you think you have substantial knowledge. Select one examination, or more if you like, and carefully read at least one of the textbooks listed in the bibliography for the test. By doing this, you will get a better idea of how much you know of what is usually taught in a college-level course in that subject. Study as much material as you can, until you think you have a good grasp of the subject matter. Then take the test at a college in your area. It will be several weeks before you receive your results, and you may wish to begin reviewing for another test in the meantime.

To find out if you are eligible for credit for your CLEP score, you must compare your score with the score required by the college you plan to attend. If you are not yet sure which college you will attend, or whether you will enroll in college at all, you should begin to follow the steps outlined. It is best that you do this before taking a CLEP test, but if you are taking the test only for the experience and to familiarize yourself with college-level material and requirements, you might take the test before you approach a college. Even if the college you decide to attend does not accept the test you took, the experience of taking such a test will enable you to meet with greater confidence the requirements of courses you will take.

You will find information about how to interpret your scores in WHAT YOUR SCORES MEAN, a booklet you will receive with your score report. Many colleges follow the recommendations of the American Council on Education (ACE) for setting their required scores, so you can use this information as a guide in determining how well you did. The ACE recommendations are included in the booklet.

If you do not do well enough on the test to earn college credit, don't be discouraged. Usually, it is the best college students who are exempted from courses or receive credit-by-examination. The fact that you cannot get credit for your score means that you should probably enroll in a college course to learn the material. However, if your score was close to the required score, or if you feel you could do better on a second try or after some additional study, you may retake the test after six months. Do not take it sooner or your score will not be reported and your fee will be forfeited.

If you do earn the score required to earn credit, you will have demonstrated that you already have some college-level knowledge. You will also have a better idea whether you should take additional CLEP examinations. And, what is most important, you can enroll in college with confidence, knowing that you do have the ability to succeed.

PREPARING TO TAKE CLEP EXAMINATIONS

Having made the decision to take one or more CLEP examinations, most people then want to know if it is worthwhile to prepare for them - how much, how long, when, and how should they go about it? The precise answers to these questions vary greatly from individual to individual. However, most candidates find that some type of test preparation is helpful.

Most people who take CLEP examinations do so to show that they have already learned the important material that is taught in a college course. Many of them need only a quick review to assure themselves that they have not forgotten some of what they once studied, and to fill in some of the gaps in their knowledge of the subject. Others feel that they need a thorough review and spend several weeks studying for a test. A few wish to take a CLEP examination as a kind of *final examination* for independent study of a subject instead of the college course. This last group requires significantly more study than those who only need to review, and they may need some guidance from professors of the subjects they are studying.

The key to how you prepare for CLEP examinations often lies in locating those skills and areas of prior learning in which you are strong and deciding where to focus your energies. Some people may know a great deal about a certain subject area, but may not test well. These individuals would probably be just as concerned about strengthening their test-taking skills as they are about studying for a specific test. Many mental and physical skills are used in preparing for a test. It is important not only to review or study for the examinations, but to make certain that you are alert, relatively free of anxiety, and aware of how to approach standardized tests. Suggestions on developing test-taking skills and preparing psychologically and physically for a test are given. The following section suggests ways of assessing your knowledge of the content of a test and then reviewing and studying the material.

Using This Study Guide

Begin by carefully reading the test description and outline of knowledge and skills required for the examination, if given. As you read through the topics listed there, ask yourself how much you know about each one. Also note the terms, names, and symbols that are mentioned, and ask yourself whether you are familiar with them. This will give you a quick overview of how much you know about the subject. If you are familiar with nearly all the material, you will probably need a minimum of review; however, if less than half of it is familiar, you will probably require substantial study to do well on the test.

If, after reviewing the test description, you find that you need extensive review, delay answering the sample question until you have done some reading in the subject. If you complete them before reviewing the material, you will probably look for the answers as you study, and then they will not be a good assessment of your ability at a later date.

If you think you are familiar with most of the test material, try to answer the sample questions.

Apply the test-taking strategies given. Keeping within the time limit suggested will give you a rough idea of how quickly you should work in order to complete the actual test.

Check your answers against the answer key. If you answered nearly all the questions correctly, you probably do not need to study the subject extensively. If you got about half the questions correct, you ought to review at least one textbook or other suggested materials on the subject. If you answered less than half the questions correctly, you will probably benefit from more extensive reading in the subject and thorough study of one or more textbooks. The textbooks listed are used at many colleges but they are not the only good texts. You will find helpful almost any standard text available to you, such as the textbook used at your college, or earlier editions of texts listed. For some examinations, topic outlines and textbooks may not be available. Take the sample tests in this book and check your answers at the end of each test. Check wrong answers.

Suggestions for Studying

The following suggestions have been gathered from people who have prepared for CLEP examinations or other college-level tests.

1. Define your goals and locate study materials.

First, determine your study goals. Set aside a block of time to review the material provided in this book, and then decide which test(s) you will take. Using the suggestions, locate suitable resource materials. If a preparation course is offered by an adult school or college in your area, you might find it helpful to enroll.

2. Find a good place to study.

To determine what kind of place you need for studying, ask yourself questions such as, *Do I need a quiet place? Does the telephone distract me? Do objects I see in this place remind me of things I should do? Is it too warm? Is it well lit? Am I too comfortable here? Do I have space to spread out my materials?* You may find the library more conducive to studying than your home. If you decide to study at home, you might prevent interruptions by other household members by putting a sign on the door of your study room to indicate when you will be available.

3. Schedule time to study.

To help you determine where studying best fits into your schedule, try this exercise: Make a list of your daily activities (for example, sleeping, working, and eating) and estimate how many hours per day you spend on each activity. Now, rate all the activities on your list in order of their importance and evaluate your use of time. Often people are astonished at how an average day appears from this perspective. They may discover that they were unaware how large portions of time are spent, or they learn their time can be scheduled in alternative ways. For example, they can remove the least important activities from their day and devote that time to studying or another important activity.

4. Establish a study routine and a set of goals.

In order to study effectively, you should establish specific goals and a schedule for accomplishing them. Some people find it helpful to write out a weekly schedule and cross out each study period when it is completed. Others maintain their concentration better by writing down the time when they expect to complete a study task. Most people find short periods of intense study more productive than long stretches of time. For example, they may follow a regular schedule of several 20- or 30-minute study periods with short breaks between them. Some people like to allow themselves rewards as they complete each study goal. It is not essential that you accomplish every goal exactly within your schedule; the point is to be committed to your task.

5. Learn how to take an active role in studying.

If you have not done much studying for some time, you may find it difficult to concentrate at first. Try a method of studying, such as the one outlined below, that will help you concentrate on and remember what you read.

a. First, read the chapter summary and the introduction. Then you will know what to look for in your reading.

b. Next, convert the section or paragraph headlines into questions. For example, if you are reading a section entitled, *The Causes of the American Revolution*, ask yourself, *What were the causes of the American Revolution?* Compose the answer as you read the paragraph. Reading and answering questions aloud will help you understand and remember the material.

c. Take notes on key ideas or concepts as you read. Writing will also help you fix concepts more firmly in your mind. Underlining key ideas or writing notes in your book can be helpful and will be useful for review. Underline only important points. If you underline more than a third of each paragraph, you are probably underlining too much.

d. If there are questions or problems at the end of a chapter, answer or solve them on paper as if you were asked to do them for homework. Mathematics textbooks (and some other books) sometimes include answers to some or all of the exercises. If you have such a book, write your answers before looking at the ones given. When problem-solving is involved, work enough problems to master the required methods and concepts. If you have difficulty with problems, review any sample problems or explanations in the chapter.

e. To retain knowledge, most people have to review the material periodically. If you are preparing for a test over an extended period of time, review key concepts and notes each week or so. Do not wait for weeks to review the material or you will need to relearn much of it.

Psychological and Physical Preparation

Most people feel at least some nervousness before taking a test. Adults who are returning to college may not have taken a test in many years or they may have had little experience with standardized tests. Some younger students, as well, are uncomfortable with testing situations. People who received their education in countries outside the United States may find that many tests given in this country are quite different from the ones they are accustomed to taking.

Not only might candidates find the types of tests and the kinds of questions on them unfamiliar, but other aspects of the testing environment may be strange as well. The physical and mental stress that results from meeting this new experience can hinder a candidate's ability to demonstrate his or her true degree of knowledge in the subject area being tested. For this reason, it is important to go to the test center well prepared, both mentally and physically, for taking the test. You may find the following suggestions helpful.

1. Familiarize yourself, as much as possible, with the test and the test situation before the day of the examination. It will be helpful for you to know ahead of time:

    a. How much time will be allowed for the test and whether there are timed subsections.

    b. What types of questions and directions appear on the examination.

    c. How your test score will be computed.

    d. Whether you will be expected to record your answers on an answer sheet or on a microcomputer.

    e. How to complete the answer sheet.

    f. How a microcomputer looks and operates, if you will be using one. (Simple instructions will be given on the day of the test, and you will be given ample time to use the computer in answering practice questions before you begin the actual test.)

    g. In which building and room the examination will be administered If you don't know where the building is, locate it or get directions ahead of time.

    h. The time of the test administration. You might wish to confirm this information a day or two before the examination and find out what time the building and room will be open so that you can plan to arrive early.

    i. Where to park your car or, if you wish to take public transportation, which bus or train to take and the location of the nearest stop.

j. Whether smoking will be permitted during the test.

k. Whether there will be a break between examinations (if you will be taking more than one on the same day), and whether there is a place nearby where you can get something to eat or drink.

2. Go to the test situation relaxed and alert. In order to prepare for the test:

   a. Get a good night's sleep. Last minute cramming, particularly late the night before, is usually counterproductive.

   b. Eat normally. It is usually not wise to skip breakfast or lunch on the day of the test or to eat a big meal just before the test.

   c. Avoid tranquilizers and stimulants. If you follow the other directions in this book, you won't need artificial aids. It's better to be a little tense than to be drowsy, but stimulants such as coffee and cola can make you nervous and interfere with your concentration.

   d. Don't drink a lot of liquids before the test. Having to leave the room during the test will disturb your concentration and take valuable time away from the test.

   e. If you are inclined to be nervous or tense, learn some relaxation exercises and use them before and perhaps during the test.

3. Arrive for the test early and prepared. Be sure to:

   a. Arrive early enough so that you can find a parking place, locate the test center, and get settled comfortably before testing begins. Allow some extra time in case you are delayed unexpectedly.

   b. Take the following with you:

      - Your completed Admission Form, which will have been validated in advance by the test administrator. You will need to refer to this form to record on your answer sheet the code number of the college to which you want your scores sent.

      - Personal identification that includes a photograph and signature. You may be asked to show such identification to be admitted.

      - Two or more soft-lead (No. 2) pencils with good erasers.

      - An unprogrammed hand calculator if you wish to use one for the Introductory Accounting or General Chemistry examination. Use of a calculator is permitted for these two tests only; it is optional and none will be supplied by the test center. Candidates are not permitted to share calculators or other supplies.

- A watch so that you can time your progress.

- Your glasses if you need them for reading or seeing the chalkboard or wall clock.

c. Leave all books, papers, and notes outside the test center. You will not be permitted to use your own scratch paper, but there will be adequate space for notes and calculations on the pages of your test book.

d. Be prepared for any temperature in the testing room. Wear layers of clothing that can be removed if the room is too hot but will keep you warm if it is too cold.

4. When you enter the test room:

a. Sit in a seat that provides a maximum of comfort and freedom from distraction. If you are bothered by cigarette smoke and smoking is allowed, ask for a seat away from the smokers. If you are lefthanded, avoid being assigned to a seat with a righthanded writing board, if possible.

b. Read directions carefully, and listen to all instructions given by the test administrator. If you don't understand the directions, ask for help before test timing begins. If you must ask a question after the test has begun, raise your hand and a proctor will assist you. The proctor can answer certain kinds of questions but cannot help you with the test.

c. Know your rights as a test taker. You can expect to be given the full working time allowed for the test(s) and a reasonably quiet and comfortable place in which to work. If a poor test situation is preventing you from doing your best, ask if the situation can be remedied. If bad test conditions cannot be remedied, ask the person in charge to report the problem in the Irregularity Report that will be sent to ETS with the answer sheets. You may also wish to immediately write a letter to CLEP, P.O. Box 6601, Princeton, NJ 08541-6601. Describe the exact circumstances as completely as you can. Be sure to include the test date and name(s) of the test(s) you took. ETS will investigate the problem to make sure it does not happen again, and, if the problem is serious enough, may arrange for you to retake the test without charge.

d. If you use an answer sheet, complete it correctly. CLEP answer sheets are scored by an electronic optical scanner that reads the marks you make on your answer sheet. Therefore, it is important that all the marks you put on your answer sheet are dark enough, that you put them in the right places, and that you erase completely all marks you do not want the scanner to read. Use a soft-lead (No. 2) pencil when completing your answer sheet.

TAKING THE EXAMINATIONS

A person may know a great deal about the subject being tested, but not do as well as he or she is capable of on the test. Knowing how to approach a test is an important part of the testing process. While a command of test-taking skills cannot substitute for knowledge of the subject matter, it can be a significant factor in successful testing.

Test-taking skills enable a person to use all available information to earn a score that truly reflects his or her ability. There are different strategies for approaching different kinds of test questions. For example, free-response questions require a very different tack than do multiple-choice questions. Other factors, such as how the test will be graded, may also influence your approach to the test and your use of test time. Thus, your preparation for a test should include finding out all you can about the test so that you can use the most effective test-taking strategies.

Before taking a test, you should know approximately how many questions are on the test, how much time you will be allowed, how the test will be scored or graded, whether there is a penalty for wrong answers, what types of questions and directions are on the test, and how you will be required to record your answers.

Taking Multiple-Choice Tests

1. Listen carefully to the instructions given by the test administrator and read carefully all directions in the test book before you begin to answer the questions.

2. Note the time that the test administrator starts timing the test. As you proceed, make sure that you are not working too slowly. You should have answered at least half the questions in a section when half the time for that section has passed. If you have not reached that point in the section, speed up your pace on the remaining questions.

3. Before answering a question, read the entire question, including all the answer choices. Don't think that because the first or second answer choice looks good to you, it isn't necessary to read the remaining options. Instructions usually tell you to select the *best* answer. Sometimes one answer choice is partially correct, but another option is better; therefore, it is usually a good idea to read all the answers before you choose one.

4. Read and consider every question. Questions that look complicated at first glance may not actually be so difficult once you have read them carefully.

5. Do not puzzle too long over any one question. If you don't know the answer after you've considered it briefly, go on to the next question. Mark that question in your test book, and go back to it later, if you have time.

6. Make sure you record your response on the answer sheet or microcomputer screen beside the same number as the number of the question in the test book. If you skip a question, be sure you also skip a space on the answer sheet or microcomputer screen. If you have not answered all the questions in a section when you are told to go on to the next section, be sure to skip the spaces for those questions on your answer sheet. If you discover that you have been marking your answers in the wrong spaces, tell the test supervisor. He or she will note this in the Irregularity Report that will be returned to ETS with the answer sheets, and your answer sheet will be hand-scored.

7. Certain foreign language and mathematics tests have only four answer options to choose from, even though the answer sheet has five lettered answer spaces for each question. Be careful not to mark answers you intend as *D* in the *E* spaces; *E* answers will not be counted toward your score.

8. Don't hesitate to write notes or do your calculations in your test book. The test books will not be used again, nor will anything you write in the test book affect your score. Do not, however, make any unnecessary marks on your answer sheet, and be sure you erase all stray marks completely.

9. In trying to determine the correct answer, you may find it helpful to cross out those options that you know are incorrect, and to make marks next to those you think might be correct. If you decide to skip the question and come back to it later, you will save yourself the time of reconsidering all the options.

10. Watch for the following key words in test questions:

all	generally	never	perhaps
always	however	none	rarely
but	may	not	seldom
except	must	often	sometimes
every	necessary	only	usually

When a question or answer option contains words such as *always*, *every*, *only*, *never*, and *none*, there can be no exceptions to the answer you choose. Use of words such as *often*, *rarely*, *sometimes*, and *generally* indicates that there may be some exceptions to the answer.

11. Do not guess wildly or randomly fill in answers to questions you do not have time to read, because there is a penalty for incorrect answers on CLEP examinations. It is improbable that mere guessing will improve your score significantly; it may even lower your score, and it does take time. If, however, you are not sure of the correct answer but have some knowledge of the question and are able to elimi-

nate one or more of the answer choices as wrong, your chance of getting the right answer is improved, and, on the average, it will be to your benefit to answer such a question.

12. Do not waste your time looking for clues to right answers based on flaws in question wording or patterns in correct answers. Professionals at the College Board and ETS put a great deal of effort into developing valid, reliable, fair tests. CLEP test development committees are composed of college faculty who are experts in the subject covered by the test and are appointed by the College Board to write test questions and to scrutinize each question that is included on a CLEP test. Committee members make every effort to ensure that the questions are not ambiguous, that they have only one correct answer, and that they cover college-level topics. These committees do not intentionally include *trick* questions. If you think a question is flawed, ask the test administrator to report it, or write immediately to College Board Test Development, P.O. Box 6600, Princeton, NJ 08541-6600. All such inquiries are investigated by the ETS test development staff.

Taking Free-Response or Essay Tests

If your college requires the optional free-response or essay portion of a CLEP Subject Examination or the General Examination in English Composition with Essay, you should do some additional preparation for your CLEP test. Taking an essay or a problem-solving test is very different from taking a multiple-choice test, so you will need to use some other strategies.

The essay written as part of the General Examination in English Composition is graded by English professors from a variety of colleges and universities. A process called holistic scoring is used to rate your writing ability.

The optional free-response portions of the Subject Examinations, on the other hand, are graded by the faculty of the college you designate as a score recipient. These optional sections require candidates to compose essay responses or to solve quantitative problems. Guidelines and criteria for grading essays are not specified by the College Board or ETS. You may find it helpful, therefore, to talk with someone at your college to find out what criteria will be used to determine whether you will get credit. If the test requires essay responses, ask how much emphasis will be placed on your writing ability and your ability to organize your thoughts as opposed to your knowledge of subject matter. Find out how much weight will be given to your multiple-choice test score in comparison with your free-response grade in determining whether you will get credit. This will give you an idea where you should expend the greatest effort in preparing for and taking the test.

Here are some strategies you will find useful in taking any essay test:

1. Before you begin to write, read all questions carefully and take a few minutes to jot down some ideas you might include in each answer.

2. If you are given a choice of questions to answer, choose the questions you think you can answer most clearly and knowledgeably.

3. Determine in what order you will answer the questions. Answer those you find the easiest first so that any extra time can be spent on the more difficult questions.

4. When you know which questions you will answer and in what order, determine how much testing time remains and estimate how many minutes you will devote to each question. Unless suggested times are given for the questions or one question appears to require more or less time than the others, allot an equal amount of time to each question.

5. Before answering each question, indicate the number of the question as it is given in the test book. You need not copy the entire question from the question sheet, but it will be helpful to you and to the person grading your test if you indicate briefly the topic you are addressing - particularly if you are not answering the questions in the order in which they appear on the test.

6. Before answering each question, read it again carefully to make sure you are interpreting it correctly. Underline key words, such as those listed below, that often appear in free-response questions. Be sure you know the exact meaning of these words before taking the test.

analyze	demonstrate	enumerate	list
apply	derive	explain	outline
assess	describe	generalize	prove
compare	determine	illustrate	rank
contrast	discuss	interpret	show
define	distinguish	justify	summarize

If a question asks you to *outline*, *define*, or *summarize*, do not write a detailed explanation; if a question asks you to *analyze*, *explain*, *illustrate*, *interpret*, or *show*, you must do more than briefly describe the topic.

7. For problem-solving questions, such as those on some mathematics, science, and business tests, be sure to explain each step clearly so that your reasoning process is apparent. Even if your final answer is incorrect because of an error in calculation, you may get considerable partial credit if the correct reasoning is demonstrated. If, after working a problem, the final answer seems wrong, state your reasons for thinking so.

INTERPRETING YOUR SCORES

To determine if you are eligible for credit for your CLEP scores, you should refer to the policy of the college you will be attending. The policy will state either the score or the percentile that is required to earn credit at that institution. Many colleges award credit at the score levels recommended by the American Council on Education. Many others use the 50th percentile. Some require scores that are higher or lower than these.

The booklet WHAT YOUR SCORES MEAN, which you will receive with your score report, gives detailed information about interpreting your scores. A brief explanation is also given below.

How CLEP Scores Are Computed

Your score report will show the total score for each CLEP examination you take. For the General Examinations, total scores fall between 200 and 800, and, for the Subject Examinations, between 20 and 80. For most Subject Examinations, percentile ranks for total scores will be shown on the score report. For some examinations, subscores are reported; they indicate performance on subsections of certain CLEP examinations and indicate a candidate's strengths and possible weaknesses. Subscores for both General and Subject Examinations fall between 20 and 80, and percentiles are not reported for them.

When your scores are computed, you receive one point for each correct answer, and a percentage of the number of your incorrect answers is subtracted from this total to discourage random guessing. As a penalty for guessing, one-fourth of a point is subtracted for a five-choice question you answer incorrectly; one-third of a point is deducted for each incorrect answer to a four-choice question. There is no deduction for a question you do not answer. The resulting figure, called the formula score, is converted to a scaled score between 200 and 800 or between 20 and 80. This conversion makes scores earned on different forms and editions of the same CLEP examination comparable to one another.

How Free-Response Answers and Essays Are Graded

Except for the General Examination in English Composition with Essay, all essays and other free-response answers are evaluated and graded by the colleges that require them, rather than by the College Board. If you take an optional free-response section, your answers will be sent with a copy of your score report to the institution you designate on your answer sheet. If you would prefer not to have your score sent to a college until after you have seen it, your free-response answers can still be sent to the college of your choice if you request a transcript within 90 days after you take the test.

Copies of free-response answers are not held by ETS beyond 90 days, or after they have been sent to an institution.

The College Board employs college English professors to grade the essays written for the General Examination in English Composition. Each essay is read and graded by two professors, the sum of the two grades is combined with the multiple-choice score, and the result is reported as a scaled score between 200 and 800. CLEP does not report separate scores for the multiple-choice and essay sections of the General Examination in English Composition because a 45-minute test is not considered extensive enough to assess reliably a candidate's writing skills for college credit. Although the format of the two sections is very different, both measure skills required for expository writing. Knowledge of formal grammar is necessary for the multiple-choice section, but the emphasis is on writing skills rather than grammar.

Notes

APPENDIX A 1-12

For a current listing of CLEP Colleges

listing where you can get credit and where you can be tested write:

CLEP, P.O. Box 6600 Princeton, N.J. 08541-6600

Or e-mail: clep @ ets.org or call: 1-609-771-7865

General Chemistry

DESCRIPTION OF THE TEST

The General Chemistry examination covers material usually taught in a one-year introductory course in general chemistry. Understanding of the structure and states of matter, reaction types, equations and stoichiometry, equilibrium, kinetics, thermodynamics, and descriptive and experimental chemistry is required, as is the ability to interpret and apply this material to new and unfamiliar problems.

Battery-operated hand-held calculators may be used during the examination; however, all calculator memories must be cleared of both programs and data and no peripheral devices such as magnetic cards or tapes are permitted. It may be helpful to use a calculator for some questions on the examination. A Periodic Chart of the elements is included in the text booklet.

The test contains approximately 80 multiple-choice questions to be answered within two separately timed 45-minute periods.

KNOWLEDGE AND SKILLS REQUIRED

Questions on the test require candidates to demonstrate one or more of the following abilities:

- Recall: remember specific facts; demonstrate straightforward knowledge of information and familiarity with terminology

- Application: understand concepts and reformulate information into other equivalent terms; apply knowledge to unfamiliar and/or practical situations; solve mathematical problems

- Interpretation: infer and deduce from data available and integrate information to form conclusions; recognize unstated assumptions

The subject matter of the General Chemistry examination is drawn from the following topics.

Approximate Percent of Examination

20% Structure of Matter

- Atomic theory and atomic structure
  - Evidence for the atomic theory
  - Atomic weights; determination by chemical and physical means
  - Atomic number and mass number; isotopes and mass spectroscopy
  - Electron energy levels; atomic spectra, quantum numbers, atomic orbitals
  - Periodic relationships including, for example, atomic radii, ionization energies, electron affinities, oxidation states

- Chemical bonding
  - Binding forces
    - Types: covalent, ionic, metallic, macromolecular (or network), van der Waals, hydrogen bonding
    - Relationships to structure and to properties
    - Polarity of bonds, electronegativities
  - Geometry of molecules, ions, and coordination complexes; structural isomerism; dipole moments of molecules; relation of properties to structure
  - Molecular models
    - Valence bond theory; hybridization of orbitals, resonance, sigma and pi bonds
    - Other models, for example, molecular orbital
- Nuclear chemistry; nuclear equations, half-lives, and radioactivity; chemical applications

19% States of Matter

- Gases
  - Laws of ideal gases; equations of state for an ideal gas
  - Kinetic-molecular theory
    - Interpretation of ideal gas laws on the basis of this theory
    - Dependence of kinetic energy of molecules on temperature; Boltzmann distribution
    - Deviations from ideal gas laws
- Liquids and solids
  - Liquids and solids from the kinetic-molecular viewpoint
  - Phase diagrams of one-component systems
  - Changes of state, critical phenomena
  - Crystal structure
- Solutions
  - Types of solutions and factors affecting solubility
  - Methods of expressing concentration
  - Colligative properties, for example, Raoult's law
  - Effect of interionic attraction on colligative properties and solubility

12% Reaction Types

- Formation and cleavage of covalent bonds
  - Acid-base reactions; concepts of Arrhenius, Bronsted-Lowry, and Lewis; amphoterism
  - Reactions involving coordination complexes
- Precipitation reactions
- Oxidation-reduction reactions
  - Oxidation number
  - The role of the electron in oxidation-reduction
  - Electrochemistry; electrolytic cells, standard half-cell potentials, prediction of the direction of redox reactions, effect of concentration changes

Approximate Percent of Examination

10% Equations and Stoichiometry

- The mole concept; Avogadro's number
- Ionic and molecular species present in chemical systems; net ionic equations
- Stoichiometry: weight and volume relations with emphasis on the mole concept
- Balancing of equations including those for redox reactions

7% Equilibrium

- Concept of dynamic equilibrium, physical and chemical; LeChatelier's principle; equilibrium constants
- Quantitative treatment
  - Equilibrium constants for gaseous reactions in terms of both molar concentrations and partial pressure (K_c, K_p)
  - Equilibrium constants for reactions in solutions
    - Constants for acids and bases; pK; pH
    - Solubility product constants and their application to precipitation and the dissolution of slightly soluble compounds
    - Constants for complex ions
    - Common ion effect; buffers

4% Kinetics

- Concept of rate of reaction
- Order of reaction and rate constant: their determination from experimental data
- Effect of temperature change on rates
- Energy of activation; the role of catalysts
- The relationship between the rate-determining step and a mechanism

5% Thermodynamics

- State functions
- First law: heat of formation; heat of reaction; change in enthalpy, Hess's law; heat capacity; heats of vaporization and fusion
- Second law: free energy of formation; free energy of reaction; dependence of change in free energy on enthalpy and entropy changes
- Relationship of change in free energy to equilibrium constants and electrode potentials

14% Descriptive Chemistry

- The accumulation of certain specific facts of chemistry is essential to enable students to comprehend the development of principles and concepts, to demonstrate applications of principle, to relate fact to theory and properties to structure, and to develop an understanding of systematic nomenclature, which facilitates communication. The following areas are normally included on the examination:

Chemical reactivity and products of chemical reactions
Chemistry of the main groups and transition elements, including typical examples of each
Organic chemistry, including such topics as functional groups and isomerism (may be treated as a separate unit or as exemplary material in other areas such as bonding)

9% Experimental Chemistry

- Some questions are based on laboratory experiments widely performed in general chemistry and ask about the equipment used, observations made, calculations performed, and interpretation of the results. The questions are designed to provide a measure of students' understanding of the basic tools of chemistry and their applications to simple chemical systems.

SAMPLE QUESTIONS

The 28 sample questions that follow are similar to questions on the General Chemistry examination, but they do not actually appear on the examination.

Before attempting to answer the sample questions, read all the information about the General Chemistry examination given above.

Try to answer correctly as many questions as possible within 30 minutes. Then compare your answers with the correct answers on the last page.

Directions: The group of questions below consists of five lettered headings followed by a list of numbered phrases. For each numbered phrase, select the one heading that is most closely related to it. Each heading may be used once, more than once, or not at all.

Questions 1-3.

A. Trigonal pyramidal
B. Tetrahedral
C. Trigonal bipyramidal
D. Square pyramidal
E. Square planar

1. The shape of the PCl_5 molecule in the vapor state 1.___

2. The shape of NH_4^+ 2.___

3. The shape of XeF_4 3.___

Directions: Each of the questions or incomplete statements below is followed by five suggested answers or completions. Select the one that is BEST in each case.

4. $1s^2 2s^2 2p^6 3s^2 3p^3$ 4.___
An atom of phosphorus has the electronic configuration indicated above. In the ground state, the number of orbitals occupied by at least one electron is
A. 3 B. 5 C. 9 D. 11 E. 15

5. For which of the following pairs of species is the difference in radius the greatest? 5.___
A. Y^{3+} and La^{3+}
B. Fe^{2+} and Co^{2+}
C. S^{2-} and Cl^-
D. Na^+ and Mg^{2+}
E. F^- and Cl^-

6. The half-life of ^{55}Cr is 1.8 hours. The delivery of a sample of this isotope from the reactor to your laboratory requires about 10.8 hours. 6.___
What is the minimum amount of such material that should be shipped in order that you receive 1.0 milligram of ^{55}Cr?
_____ mg.
A. 128 B. 64 C. 32 D. 11 E. 1.0

7. At constant temperature, the behavior of a sample of a real gas more closely approximates that of an ideal gas as its volume is increased because the 7.___
A. collisions with the walls of the container become less frequent
B. average molecular speed decreases
C. molecules have expanded
D. average distance between molecules becomes greater
E. average molecular kinetic energy decreases

8. A sealed vessel contains 0.200 mole of oxygen gas, 0.100 mole of nitrogen gas, and 0.200 mole of argon gas. The total pressure of the gas mixture is 5.00 atmospheres. The partial pressure of the argon is _____ atmosphere(s). 8.___
A. 0.200 B. 0.500 C. 1.00 D. 2.00 E. 5.00

9. The critical temperature of a substance gives information about all of the following EXCEPT the 9.___
A. slope of the solid-liquid line on a phase diagram
B. relative attractive forces among the molecules
C. highest boiling point the liquid can exhibit
D. ease of liquefying a gas by pressure
E. lowest temperature above which only one phase can exist

10. For a solid solute that has an endothermic heat of solution when dissolved in water, which of the following is TRUE? 10.___
A. Its lattice energy is less than its energy of hydration.
B. Its solubility will increase with an increase in temperature.

C. Its solubility increases significantly with an increase in pressure.
D. The value for ΔG for the solution process is positive.
E. The entropy of the system decreases as the solute is dissolved.

11. A 2.6-gram sample of a nonvolatile solute (molecular weight 52) is dissolved in 50. grams of a solvent (molecular weight 100.) which has a vapor pressure of 33 millimeters of mercury at 20°C. Assuming the solution to be ideal, the vapor pressure of the solution at 20°C is _____ mm Hg. 11.___

A. 3.0 B. 20. C. 23 D. 27 E. 30.

12. A solution prepared by mixing 10 milliliters of 1-molar HCl and 10 milliliters of 2-molar NaOH has a pH of 12.___

A. 0.0 B. 0.3 C. 7.0 D. 13.7 E. 14.0

13. All of the following reactions can be defined as Lewis acid-base reactions EXCEPT 13.___

A. $Al(OH)_3 + OH^- \rightarrow Al(OH)_4^-$
B. $Cl_2 + H_2O \rightarrow HOCl + H^+ + Cl^-$
C. $SnCl_4 + 2Cl^- \rightarrow SnCl_6^{2-}$
D. $NH_4^+ + NH_2^- \rightarrow 2NH_3$
E. $H^+ + NH_3 \rightarrow NH_4^+$

14. Which of the species below contains a group V element in an oxidation state different from that of the others? 14.___

A. NO_3^- B. $POCl_3$ C. $H_2PO_4^-$ D. $KAsF_4$ E. Sb_4O_{10}

15. 15.___

$Cd^{2+} + 2e^- \rightleftarrows Cd$ $E^0 = -0.40$ volt
$Cu^+ + e^- \rightleftarrows Cu$ $E^0 = +0.52$ volt
$Ag^+ + e^- \rightleftarrows Ag$ $E^0 = +0.80$ volt

Based on the standard electrode potentials given above, which of the following is the strongest reducing agent?

A. Cd B. Cd^{2+} C. Cu D. Ag E. Ag^+

16. An impure sample of $CaCO_3$ (molecular weight 100.) was reported as being 28.0 percent Ca. If no calcium was present as an impurity, the percent of $CaCO_3$ in the sample is 16.___

A. 28.0% B. 42.0% C. 56.0% D 70.0% E. 100.0%

17. $2Al(s) + 6HCl(aq) \rightarrow 2AlCl_3(aq) + 3H_2(g)$ 17.___

According to the above reaction at 25°C and 1.00 atmosphere, about how many grams of aluminum (atomic weight 27) are necessary to produce 0.60 mole of hydrogen gas?

A. 1.2 B. 11 C. 16 D. 24 E. 32

18. $...IO_3^- + ...S_2O_3^{2-} + ...H^+ \rightarrow ...I^- + ...S_4O_6^{2-} + ...H_2O$ 18.___

When the equation above is balanced and all coefficients are reduced to lowest whole number terms, the coefficient for the reducing agent is

A. 1 B. 3 C. 4 D. 6 E. 8

19. The ionization constant for acetic acid is 1.8×10^{-5}; that for hydrocyanic acid is 4×10^{-10}. 19.___
In 0.1-molar solutions of sodium acetate and sodium cyanide, it is TRUE that
   A. $[H^+]$ equals $[OH^-]$ in each solution
   B. $[H^+]$ exceeds $[OH^-]$ in each solution
   C. $[H^+]$ of the sodium acetate solution is less than that of the sodium cyanide solution
   D. $[OH^-]$ of the sodium acetate solution is less than that of the sodium cyanide solution
   E. $[OH^-]$ for the two solutions is the same

20. $HCl > HC_2H_3O_2 > HCN > H_2O > NH_3$ 20.___
Five acids are listed above in the order of decreasing acid strength.
Which of the following reactions must have an equilibrium constant with a value less than 1?
   A. $HCl + CN^- \rightleftarrows HCN + Cl^-$
   B. $HCl + H_2O \rightleftarrows H_3O^+ + Cl^-$
   C. $HC_2H_3O_2 + OH^- \rightleftarrows C_2H_3O_2^- + H_2O$
   D. $H_2O + NH_2^- \rightleftarrows NH_3 + OH^-$
   E. $HCN + C_2H_3O_2^- \rightleftarrows HC_2H_3O_2 + CN^-$

21. $2NO + Cl_2 \rightarrow 2NOCl$ 21.___
Doubling the concentration of both reactants in the reaction above increases the rate by a factor of 8. Doubling the chlorine concentration alone doubles the rate.
What is the overall order of the reaction?
   A. 0   B. 1   C. 2   D. 3   E. 4

22. All of the following statements about the nitrogen family of elements are true EXCEPT: 22.___
   A. It contains both metals and nonmetals
   B. The electronic configuration of the valence shell of the atom is ns^2np^3
   C. The only oxidation states exhibited by members of this family are -3,0,+3,+5
   D. The atomic radii increase with increasing atomic number
   E. The boiling points increase with increasing atomic number

23. $A \rightarrow X$ 23.___
The enthalpy change for the reaction above is ΔH_T.
This reaction can be broken down into a series of steps as shown in the diagram:

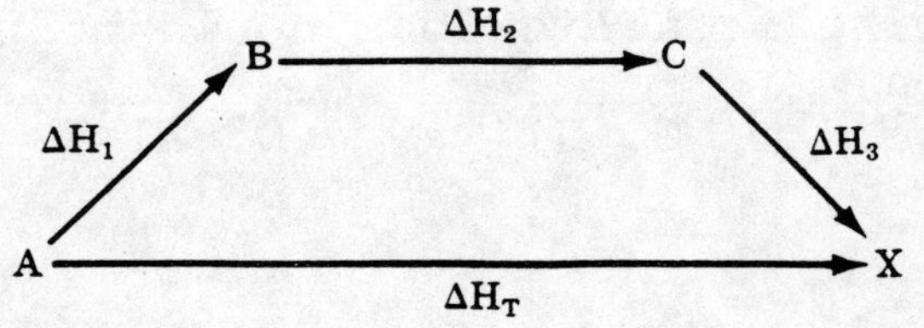

A relationship that must exist among the various enthalpy changes is

A. $\Delta H_T - \Delta H_1 - \Delta H_2 - \Delta H_3 = 0$

B. $\Delta H_T + \Delta H_1 + \Delta H_2 + \Delta H_3 = 0$

C. $\Delta H_3 - (\Delta H_1 + \Delta H_2) = \Delta H_T$

D. $\Delta H_2 - (\Delta H_3 + \Delta H_1) = \Delta H_T$

E. $\Delta H_T + \Delta H_2 = \Delta H_1 + \Delta H_3$

24. What formula would be expected for a binary compound of barium and nitrogen? 24.___

A. Ba_3N_2 B. Ba_2N_3 C. Ba_2N D. BaN_2 E. BaN

25. Which of the following salts when dissolved in water forms a basic solution? 25.___

A. NaCl B. $(NH_4)_2SO_4$ C. $CuSO_4$
D. K_2CO_3 E. NH_4NO_3

26. Of the following organic compounds, which is LEAST soluble in water? 26.___

A. CH_3OH
B. $CH_3CH_2CH_2OH$
C. $CH_3CH_2CH_2OCH_2CH_2CH_3$
D. $CH_3\overset{\overset{O}{\|}}{C}CH_3$
E. $CH_3CH_2\overset{\overset{O}{\|}}{C}OH$

27. Of the laboratory equipment listed below, which has the GREATEST relative uncertainty in a measurement? 27.___

A. 50-ml burette when used to measure 25 ml
B. 25-ml pipette when used to measure 25 ml
C. 50-ml graduated cylinder when used to measure 25 ml
D. Analytical balance when used to weigh 25 grams
E. Centigram balance when used to weigh 25 grams

28. The table below summarizes the reactions of two samples of a certain unknown solution when treated with bases. 28.___

		Results	
Sample	Reagent Used	Limited Amount of Reagent	Excess Reagent
I	NaOH	White precipitate	Precipitate dissolves
II	$NH_3(aq)$	White precipitate	White precipitate

Which of the following metallic ions could be present in the unknown solution?

A. Ca^{2+} B. Zn^{2+} C. Ni^{2+} D. Al^{3+} E. Ag^+

OPTIONAL FREE-RESPONSE SECTION

If your college requires that the optional free-response section be taken in addition to the multiple-choice examination, you should review the following information.

The optional free-response section of the General Chemistry examination requires candidates to respond to six of nine questions offered within a 90-minute period. The questions include:

- A quantitative problem based on chemical equilibria
- A quantitative problem based on applications of principles from either stoichiometry, kinetics, thermodynamics, electrochemistry, or solution chemistry
- The writing of equations that predict the formulas of the products of chemical reactions
- The writing of short essays that interpret chemical phenomena and illustrate chemical reasoning

An unprogrammed calculator may be used, and candidates are not required to show computations; however, they should show all other work.

STUDY RESOURCES

To prepare for the General Chemistry examination, study the content of one college textbook in this subject. Most first-year chemistry textbooks cover the topics in the test outline that is given on pages 2-4, but their approaches to and the emphases on certain topics may differ.

Below is a list of textbooks that are frequently used in first-year chemistry courses.

Textbooks

Brady, James E. and Gerald E. Humiston, GENERAL CHEMISTRY: PRINCIPLES AND STRUCTURE, 5th ed. New York: John Wiley and Sons, Inc., 1990.

Brown, Theodore L. and H.E. LeMay, CHEMISTRY: THE CENTRAL SCIENCE, 4th ed. Englewood Cliffs, NJ: Prentice-Hall, Inc., 1989.

Dickerson, R.E. et al., CHEMICAL PRINCIPLES, 4th ed. Menlo Park, CA: Benjamin Cummings, 1984.

Holtzclaw, Henry F. and William R. Robinson, COLLEGE CHEMISTRY WITH QUALITATIVE ANALYSIS, 8th ed. Lexington, MA: D.C. Heath and Co., 1988.

Masterton, William L. et al., CHEMICAL PRINCIPLES, 6th ed. Philadelphia: Saunders College Publications, 1985.

Petrucci, R., GENERAL CHEMISTRY, 5th ed. New York: Macmillan, 1989.

Zumdahl, S.S., CHEMISTRY, 2nd ed. Lexington, MA: D.C. Heath and Co., 1989.

KEY (CORRECT ANSWERS)

1. C	11. E	21. D
2. B	12. D	22. C
3. E	13. B	23. A
4. C	14. D	24. A
5. E	15. A	25. D
6. B	16. D	26. C
7. D	17. B	27. C
8. D	18. D	28. D
9. A	19. D	
10. B	20. E	

HOW TO TAKE A TEST

You have studied hard, long, and conscientiously.

With your official admission card in hand, and your heart pounding, you have been admitted to the examination room.

You note that there are several hundred other applicants in the examination room waiting to take the same test.

They all appear to be equally well prepared.

You know that nothing but your best effort will suffice. The "moment of truth" is at hand: you now have to demonstrate objectively, in writing, your knowledge of content and your understanding of subject matter.

You are fighting the most important battle of your life—to pass and/or score high on an examination which will determine your career and provide the economic basis for your livelihood.

What extra, special things should you know and should you do in taking the examination?

BEFORE THE TEST

YOUR PHYSICAL CONDITION IS IMPORTANT

If you are not well, you can't do your best work on tests. If you are half asleep, you can't do your best either. Here are some tips:

1. Get about the same amount of sleep you usually get. Don't stay up all night before the test, either partying or worrying—DON'T DO IT.
2. If you wear glasses, be sure to wear them when you go to take the test. This goes for hearing aids, too.
3. If you have any physical problems that may keep you from doing your best, be sure to tell the person giving the test. If you are sick or in poor health, you really cannot do your best on any test. You can always come back and take the test some other time.

AT THE TEST

EXAMINATION TECHNIQUES

1. Read the *general* instructions carefully. These are usually printed on the first page of the examination booklet. As a rule, these instructions refer to the timing of the examination; the fact that you should not start work until the signal and must stop work at a signal, etc. If there are any *special* instructions, such as a choice of questions to be answered, make sure that you note this instruction carefully.

2. When you are ready to start work on the examination, that is as soon as the signal has been given, read the instructions to each question booklet, underline any key words or phrases, such as *least, best, outline, describe*, and the like. In this way you will tend to answer as requested rather than discover on reviewing your paper that you *listed without describing*, that you selected the *worst* choice rather than the *best* choice, etc.

3. If the examination is of the objective or so-called multiple-choice type, that is, each question will also give a series of possible answers: A, B, C, or D, and you are called upon to select the best answer and write the letter next to that answer on your answer paper, it is advisable to start answering each question in turn. There may be anywhere from 50 to 100 such questions in the three or four hours allotted and you can see how much time would be taken if you read through all the questions before beginning to answer any. Furthermore, if you come across a question or a group of questions which you know would be difficult to answer, it would undoubtedly affect your handling of all the other questions.

4. If the examination is of the essay-type and contains but a few questions, it is a moot point as to whether you should read all the questions before starting to answer any one. Of course if you are given a choice, say five out of seven and the like, then it is essential to read all the questions so you can eliminate the two which are most difficult. If, however, you are asked to answer all the questions, there may be danger in trying to answer the easiest one first because you may find that you will spend too much time on it. The best technique is to answer the first question, then proceed to the second, etc.

5. Time your answers. Before the examination begins, write down the time it started, then add the time allowed for the examination and write down the time it must be completed, then divide the time available somewhat as follows:

    a. If 3½ hours are allowed, that would be 210 minutes. If you have 80 objective-type questions, that would be an average of about 2½ minutes per question. Allow yourself no more than 2 minutes per question, or a total of 160 minutes, which will permit about 50 minutes to review.

    b. If for the time allotment of 210 minutes, there are 7 essay questions to answer, that would average about 30 minutes a question. Give yourself only 25 minutes per question so that you have about 35 minutes to review.

6. The most important instruction is *to read each question* and make sure you know what is wanted. The second most important instruction is to *time yourself properly* so that you answer every question. The third most important instruction is to *answer every question*. Guess if you have to but include something for each question, Remember that you will receive no credit for a blank and will probably receive some credit if you write something in answer to an essay question. If you guess a letter, say "B" for a multiple-choice question, you may have guessed right. If you leave a blank as the answer to a multiple-choice question, the examiners may respect your feelings but it will not add a point to your score. Some exams may penalize you for wrong answers, so in such cases *only*, you may not want to guess unless you have some basis for your answer.

7. Suggestions

   a. Objective-Type Questions

      (1) Examine the question booklet for proper sequence of pages and questions.

      (2) Read all instructions carefully.

      (3) Skip any question which seems too difficult; return to it after all other questions have been answered.

      (4) Apportion your time properly; do not spend too much time on any single question or group of questions.

      (5) Note and underline key words—*all, most, fewest, least, best, worst, same, opposite*.

      (6) Pay particular attention to negatives.

      (7) Note unusual option, e.g., unduly long, short, complex, different or similar in content to the body of the question.

      (8) Observe the use of "hedging" words—*probably, may, most likely, etc.*

      (9) Make sure that your answer is put next to the same number as the question.

      (10) Do not second guess unless you have good reason to believe the second answer is definitely more correct.

      (11) Cross out original answer if you decide another answer is more accurate; do not erase, *until* you are ready to hand your paper in.

      (12) Answer all questions; guess unless instructed otherwise.

      (13) Leave time for review.

b. Essay-Type Questions

(1) Read each question carefully.

(2) Determine exactly what is wanted. Underline key words or phrases.

(3) Decide on outline or paragraph answer.

(4) Include many different points and elements unless asked to develop any one or two points or elements.

(5) Show impartiality by giving pros and cons unless directed to select one side only.

(6) Make and write down any assumptions you find necessary to answer the question.

(7) Watch your English, grammar, punctuation, choice of words.

(8) Time your answers; don't crowd material.

8. Answering the Essay Question

Most essay questions can be answered by framing the specific response around several key words or ideas. Here are a few such key words or ideas:

M's: manpower, materials, methods, money, management

P's: purpose, program, policy, plan, procedure, practice, problems, pitfalls, personnel, public relations

a. Six basic steps in handling problems:

(1) preliminary plan and background development

(2) collect information, data and facts

(3) analyze and interpret information, data and facts

(4) analyze and develop solutions as well as make recommendations

(5) prepare report and sell recommendations

(6) install recommendations and follow up effectiveness

b. Pitfalls to Avoid

(1) *Taking Things for Granted*
A statement of the situation does not necessarily imply that each of the elements is necessarily true; for example, a complaint may be invalid and biased so that all that can be taken for granted is that a complaint has been registered

(2) *Considering only one side of a situation*
Wherever possible, indicate several alternatives and then point out the reasons you selected the best one.

(3) *Failing to indicate follow up*
Whenever your answer indicates action on your part, make certain that you will take proper follow-up action to see how successful your recommendations, procedures, or actions turn out to be.

(4) *Taking too long in answering any single question*
Remember to time your answers properly.

EXAMINATION SECTION

EXAMINATION SECTION

TEST 1

DIRECTIONS: Each question or incomplete statement is followed by several suggested answers or completions. Select the one that BEST answers the question or completes the statement. *PRINT THE LETTER OF THE CORRECT ANSWER IN THE SPACE AT THE RIGHT.*

The following list of atomic weights may be referred to in solving problems involving computations:

Aluminum	27.0	Hydrogen	1.0	Phosphorous	31.0
Bromine	79.9	Iodine	126.9	Silver	107.9
Calcium	40.1	Iron	55.9	Sodium	23.0
Carbon	12.0	Lead	207.2	Sulfur	32.0
Chlorine	35.5	Nitrogen	14.0	Tin	118.7
Copper	63.5	Oxygen	16.0	Zinc	65.4

1. A solution of 0.02 M HCl has a pH of — 1. B
   A. 1.7 B. 2.0 C. 2.3 D. 2.7

2. The pH of a saturated solution of CO_2 in water at 25°C is MOST NEARLY — 2.___
   A. 3 B. 5 C. 7 D. 9

3. An aqueous solution containing 100 grams of HCl per liter of solution is APPROXIMATELY ______ N. — 2.___
   A. 1.0 B. 2.8 C. 3.2 D. 12.0

4. The volume of 12 N HCl which must be taken and made up to 100 ml with distilled water in order to make 100 ml of 1 N HCl is ______ ml. — 4.___
   A. 1.2 B. 6.5 C. 8.3 D. 12

5. Of the following, the one which has the HIGHEST melting point is — 5. D
   A. benzoic acid
   B. lead
   C. sodium chloride
   D. sulfur

6. Of the following, the one which has the LOWEST boiling point is — 6.___
   A. CH_3-CH_3
   B. CH_3-CO-CH_3
   C. CH_3-CH_2-O-CH_2-CH_3
   D. CCl_4

7. Of the following, the one which has the HIGHEST boiling point is — 7.___
   A. CH_3COOH B. HCl C. HNO_3 D. H_2SO_4

8. Of the following, the STRONGEST oxidizing agent is — 8.___
   A. $CrCl_3$ B. I_2 C. $KMnO_4$ D. $SnCl_2$

9. Of the following, the STRONGEST reducing agent is 9.___
   A. $BaCl_2$ B. $FeCl_3$ C. $Na_2S_2O_3$ D. $Na_2S_4O_6$

10. Of the following substances, the one with the HIGHEST softening point is 10.___
   A. Pyrex B. quartz
   C. soda-lime glass D. Vycor

11. The purpose of annealing glass is to 11.___
   A. give a shiny finish B. increase the ductility
   C. relieve strains D. remove fogging

12. Of the following reagents, the one which is NOT corrosive to chemical glassware is hot 12.___
   A. 12 M HCl
   B. 16 M HNO_3
   C. conc. HCl + conc. HNO_3
   D. saturated NaOH

13. Of the following reagents, the one which is MOST corrosive to a platinum crucible is 13.___
   A. a mixture of conc. HCl and conc. HNO_3
   B. boiling H_2SO_4
   C. fused NaOH
   D. hot 12 M NH_3

14. Of the following metals, the one which is MOST active chemically is 14. D
   A. copper B. iron C. mercury D. zinc

15. Of the following gases, the one which is MOST soluble in water is 15. C
   A. CO_2 B. HCl C. H_2S D. N_2

16. Of the following, the gas with the GREATEST density at standard temperature and pressure is 16.___
   A. air B. carbon dioxide
   C. hydrogen sulfide D. methane

17. Water is at its GREATEST density at a temperature of 17.___
   A. 0°C B. 4°C C. 20°C D. 37°C

18. The term Ostwald refers to a type of apparatus GENERALLY used to determine 18.___
   A. color B. refractive index
   C. temperature D. viscosity

19. The term Beilstein refers to a chemical 19.___
   A. reference book B. society
   C. technique D. unit of radioactivity

20. The term Westphal refers to a type of apparatus GENERALLY used to determine 20.___
   A. boiling point elevation B. freezing point
   C. refractive index D. specific gravity

21. The recommended solder for ordinary electrical connections is 21.___
A. Babbitt metal
B. resin-core solder
C. silver solder
D. soft solder with an acid flux

22. Soft solder is composed PRINCIPALLY of lead and 22.___
A. antimony B. bismuth C. cadmium D. tin

23. Solid Na_2CO_3 is BEST removed from a saturated solution of NaOH by 23.___
A. distillation
B. filtration through ashless filter paper
C. filtration through a Gooch crucible with asbestos
D. filtration through a sintered glass funnel

24. A Type K potentiometer is USUALLY used for measuring 24.___
A. amount of electrical charge
B. parachor
C. refractive index
D. voltage

25. A Wheatstone bridge is FREQUENTLY used to 25.___
A. calibrate photographic developer
B. determine optical rotation
C. measure electrical resistance
D. support electrodes

26. The McLeod gauge is used to measure 26.___
A. amperage B. humidity
C. low pressures D. low temperatures

27. The minimum amount of liquid needed for an accurate pyknometric determination of its density is MOST NEARLY 27.___
A. 0.1 liter B. 1.0 liter
C. 0.1 ml D. 10 ml

28. The minimum volume of liquid needed for routine measurement of its refractive index is MOST NEARLY ___ ml. 28.___
A. 0.1 B. 1.0 C. 10 D. 100

29. Of the following, the one which is NOT intended for measuring temperature is the 29.___
A. iron-contstantan thermocouple
B. mercury thermoregulator
C. nickel resistance thermometer
D. optical pyrometer

30. The only one of the following substances which is SAFE to heat is 30.___
A. ammonium nitrate B. lead azide
C. perchloric acid D. potassium persulfate

31. The PRINCIPAL function of the *fix* in photographic processing is to 31.___
   A. dissolve the unreacted gelatine
   B. dissolve the unreacted silver
   C. dissolve the unreacted silver halide
   D. tan the emulsion

32. The temperature of an incandescent object which is glowing cherry red is APPROXIMATELY _____ °C. 32.___
   A. 250 B. 400 C. 700 D. 1300

33. Chromatography is used PRINCIPALLY for 33.___
   A. carbon and hydrogen analysis of organic compounds
   B. purification of industrial water supplies
   C. quantitative inorganic analysis
   D. separation of small amounts of organic compounds from mixtures

34. A unit of radioactivity is the 34.___
   A. microcurie B. microlumen
   C. microradian D. microwave

35. Of the following substances, the one with the LOWEST solubility in water (measured in grams per liter) is 35.___
   A. barium chloride B. calcium nitrate
   C. ferrous sulfate D. mercurous chloride

36. The solubility product of AgCl is 10^{-10}. The solubility of AgCl in a 0.1 M NaCl solution is MOST NEARLY _____ moles/liter. 36.___
   A. 10^{-11} B. 10^{-9} C. 10^{-5} D. 10^{-1}

37. A violet color in a flame test indicates the presence of 37.___
   A. lithium B. potassium C. sodium D. strontium

38. The one of the following which will dissolve 10 gms of AgI is 100 ml of 38.___
   A. 1 M HNO_3 B. 1 M NH_3 C. 6 M HCl D. 6 M NaCN

39. The one of the following substances which is NOT white is 39.___
   A. aluminum hydroxide B. cuprous chloride
   C. mercuric iodide D. zinc sulfate

40. An unknown white solid dissolves in 6 M NH_3 to give a clear colorless solution. It, therefore, CANNOT contain _____ ions. 40.___
   A. magnesium B. mercuric C. silver D. zinc

KEY (CORRECT ANSWERS)

1. A	11. C	21. B	31. C
2. B	12. D	22. D	32. C
3. B	13. A	23. C	33. D
4. C	14. D	24. D	34. A
5. C	15. B	25. C	35. D
6. A	16. B	26. C	36. B
7. D	17. B	27. D	37. B
8. C	18. D	28. A	38. D
9. C	19. A	29. B	39. C
10. B	20. D	30. D	40. B

TEST 2

DIRECTIONS: Each question or incomplete statement is followed by several suggested answers or completions. Select the one that BEST answers the question or completes the statement. *PRINT THE LETTER OF THE CORRECT ANSWER IN THE SPACE AT THE RIGHT.*

1. In titrating 0.1 M acetic acid with 0.1 M NaOH, the CORRECT endpoint will be at a pH of APPROXIMATELY 1.___
   A. 2 B. 5 C. 9 D. 11

2. Potassium thiocyanate gives a reddish color with aqueous solutions of _____ ions. 2.___
   A. cuprous B. ferric C. lead D. magnesium

3. Of the following, the one used in the colorimetric test for low concentrations of ammonia in water is 3.___
   A. Fehling's Solution B. Karl Fischer Reagent
   C. Nessler Reagent D. Ringer's Solution

4. Dimethylglyoxime is a reagent used to detect the presence of 4.___
   A. chlorine B. lead C. nickel D. tin

5. In the gravimetric analysis of a chloride, the weighing form is NORMALLY 5.___
   A. AgCl B. $BaCl_2$ C. Hg_2Cl_2 D. NaCl

6. Of the following salts, the one which is MOST soluble in water at room temperature is 6.___
   A. Ag_2SO_4 B. $BaSO_4$ C. CaC_2O_4 D. $MgSO_4$

7. The one of the following solids which is MOST soluble in 8 M NaOH is 7.___
   A. $BaSO_4$
   B. $Fe(OH)_3$
   C. hydrated aluminum hydroxide
   D. hydrated silver hydroxide

8. Of the following, the one which is the amide group is 8.___
   A. $-\overset{O}{\overset{\|}{C}}-NH_2$ B. $-N\begin{matrix}\diagup CH_3 \\ \diagdown CH_3\end{matrix}$ C. $-NH_2$ D. $-N{=}O$

9. The linkage between the amino acid units of a protein is called 9.___
   A. ether linkage B. hydrogen bond
   C. oxazole linkage D. peptide linkage

10. Of the following compounds, the one which will react MOST vigorously with sodium at room temperature is 10.___

A. CH_3CH_2OH
B. $CH_3-\underset{\underset{O}{\|}}{C}-O-CH_3$
C. $CH_3CH_2-O-CH_2CH_3$
D. 

11. Of the following compounds, the one which possesses geometrical isomers is 11.___

A. $CCl_2=CCl_2$ B. $CH_2=CCl_2$ C. $CH_2=CHCl$ D. $CHCl=CHCl$

12. Of the following, the one which is the STRONGEST acid is 12.___

A. CCl_3COOH B. $CHCl_2COOH$ C. $CH_2ClCOOH$ D. CH_3COOH

13. Glucose is a 13.___

A. disaccharide
B. monosaccharide
C. polysaccharide
D. starch

14. Of the following, the one which is an example of an amino acid is 14.___

A. cellulose B. diastase C. maltose D. tyrosine

15. Of the following acids, the one with the LARGEST acid constant is 15.___

A. acetic acid
B. carbonic acid
C. hydrocyanic acid
D. hydrogen sulfide

16. A water solution of calcium oxide is 16.___

A. acid B. alkaline C. neutral D. a base

17. A water solution of aluminum chloride is 17.___

A. acid B. alkaline C. neutral D. a base

18. The vapor pressure of mercury at room temperature is APPROXIMATELY ______ mm Hg. 18.___

A. 10^{-6} B. 10^{-3} C. 1 D. 760

19. Of the following elements, the ONLY one which is ferromagnetic is 19.___

A. aluminum B. boron C. nickel D. tungsten

20. Mixing gelatine with water gives a(n) 20.___

A. emulsion
B. lyophilic colloid
C. lyophobic colloid
D. suspension

21. In the reaction $Ra_{226}^{90} \rightarrow Th_{222}^{88} + X$, X is a(n) 21.___

A. alpha particle
B. beta particle
C. gamma ray
D. meson

22. Of the following, the one with the LEAST mass is the 22.___
A. deuteron B. electron C. neutron D. proton

23. Beer's law refers to the 23.___
A. absorption of light in solutions
B. boiling point of non-associated liquids
C. surface tension of a homologous series of fatty acids
D. vapor pressure of binary liquid mixtures

24. A spectrophotometer working in the ultraviolet and visible regions of the spectrum is PARTICULARLY useful in 24.___
A. colorimetric analysis
B. distinguishing the components of gasoline
C. distinguishing optical isomers
D. qualitative analysis of inorganic salts

25. The addition of a neutron to a nucleus 25.___
A. raises the atomic number by one without changing the atomic weight
B. raises the atomic weight by one without changing atomic weight
C. raises both the atomic number and the atomic weight by one
D. always causes nuclear fission to occur

26. Of the following elements, the only one which CANNOT be tested for on an emission spectrograph is 26.___
A. copper B. iron C. mercury D. oxygen

27. A gas which is LIGHTER than air has the formula 27.___
A. C_2H_2 B. C_2H_6 C. N_2O D. NO_2

28. The kinetic theory of gases predicts that the viscosity of gases will _____ density. 28.___
A. decrease with decreased
B. vary with the square root of the
C. increase with decreased
D. be independent of

29. The specific resistance of a 0.1 N NaCl solution was found to be 93.6 ohms at 25°C. 29.___
Its SPECIFIC conductance will be _____ ohms.
A. 0.936 B. 6.4 reciprocal
C. 0.011 reciprocal D. 93.6

30. Pure carbon monoxide may be prepared in the laboratory by 30.___
A. adding dilute sulfuric acid to marble chips
B. adding concentrated sulfuric acid to formic acid
C. burning coke in a limited supply of air
D. adding dilute hydrochloric acid to limestone

31. Pound for pound, sodium carbonate, when compared with sodium bicarbonate, will provide _____ sodium ions and _____ carbon dioxide. 31.___
A. fewer; less
B. more; less
C. more; more
D. fewer; more

32. Water gas (synthesis gas) is MAINLY a mixture of 32.___
A. CO and N_2
B. CO and H_2
C. CO_2 and H_2O
D. CO and O_2

33. The anhydride of nitrous acid is 33.___
A. NO
B. N_2O_3
C. N_2O_5
D. NO_2

34. The percent of gold in 14-carat gold is APPROXIMATELY 34.___
A. 58
B. 78
C. 83
D. 70

35. Three of the following are ores of iron. The one which is NOT is 35.___
A. siderite
B. hematite
C. limonite
D. spiegeleisen

36. The symbol of the element with the GREATEST tendency to gain electrons is 36.___
A. F
B. Bi
C. At
D. O

37. The Kroll process is used for extracting 37.___
A. geramium
B. molybdenum
C. titanium
D. wolfram

38. Waste rock material in an ore is called 38.___
A. gangue
B. slag
C. flux
D. froth

39. Cementite is of fundamental importance in iron alloys. Its formula is 39.___
A. FeC
B. Fe_2C
C. Fe_3C
D. Fe_4C_3

40. Lead is dissolved MOST readily in dilute _____ acid. 40.___
A. acetic
B. sulfuric
C. hydrochloric
D. phosphoric

KEY (CORRECT ANSWERS)

1. C	11. D	21. A	31. D
2. B	12. A	22. B	32. A
3. C	13. B	23. A	33. A
4. C	14. D	24. A	34. B
5. A	15. A	25. D	35. C
6. D	16. B	26. D	36. A
7. C	17. A	27. A	37. B
8. A	18. B	28. C	38. A
9. D	19. C	29. A	39. A
10. A	20. B	30. D	40. A

EXAMINATION SECTION

TEST 1

DIRECTIONS: Each question or incomplete statement is followed by several suggested answers or completions. Select the one that BEST answers the question or completes the statement. *PRINT THE LETTER OF THE CORRECT ANSWER IN THE SPACE AT THE RIGHT.*

1. The industrial preparation of methanol from wood is an example of 1.____
   A. crystallization B. destructive distillation
   C. hydrogenation D. organic fermentation

2. An emulsion is a dispersion of a 2.____
   A. gas in a liquid B. liquid in a liquid
   C. liquid in a solid D. solid in a liquid

3. In volumetric analysis, it is NOT recommended to titrate a 3.____
   A. strong acid with a strong base
   B. weak acid with a strong base
   C. weak base with a strong acid
   D. weak base with a weak acid

4. Of the following, the element which forms an amphoteric oxide is 4.____
   A. arsenic B. chlorine C. silver D. sulfur

5. The term used to indicate water-free chemicals is 5.____
   A. abs. B. anhyd. C. C.P. D. "technical"

6. X-rays are MOST similar to 6.____
   A. alpha rays B. beta rays
   C. electrons D. gamma rays

7. Of the following atoms, the one with the LOWEST atomic weight is 7.____
   A. lead B. radium C. radon D. uranium

8. Of the following atomic particles, the one with the GREATEST mass is the 8.____
   A. alpha particle B. beta particle
   C. neutron D. proton

9. The *angstrom* is a unit of 9.____
   A. length B. mass C. volume D. weight

10. The chemical symbol for plutonium is 10.____
    A. Pn B. Pt C. Pu D. Pv

11. Of the following, the metal of GREATEST chemical activity is 11.____
    A. calcium B. copper C. mercury D. nickel

12. Of the following, the chemical which is COMMONLY used as an oxidizing agent in analytical work is 12.____
A. $H_2C_2O_4$ B. Iron C. Na_2SO_3 D. PbO_2

13. Sodium stannite is used in the confirmatory qualitative test for 13.____
A. bismuth B. calcium C. chromium D. silver

14. An acid solution is one in which the $[H^+]$ is 14.____
A. equal to 10^{-7} mole/liter
B. greater than 10^{-7} mole/liter
C. less than 10^{-7} mole/liter
D. less than 10^{-14} mole/liter

15. Beer's law is the BASIS of calculations in 15.____
A. acidimetry
B. calorimetry
C. oxidation-reduction reactions
D. photometry

16. Charles' law states: At constant 16.____
A. pressure, the volume of a given mass of gas varies directly with the absolute temperature
B. temperature, the volume of a given mass of gas varies inversely with the pressure
C. volume, the pressure of a given mass of gas varies directly with the absolute temperature
D. volume, the pressure of a given mass of gas varies inversely with the absolute temperature

17. According to LeChatelier's law, if the pressure is increased in the system $H_2 + I_2 \rightleftarrows 2\ HI$ (assume that the system is in equilibrium and that all substances are in the gaseous state), the reaction would 17.____
A. go farther to the left
B. go farther to the right
C. go farther to the center
D. not be affected

18. According to Van't Hoff's law, if the temperature is raised in the system $CO + 2\ H_2 \rightleftarrows CH_3OH + 24{,}000$ calories (assume that the system is in equilibrium and that all substances are in the gaseous state), the reaction would 18.____
A. go farther to the left
B. go farther to the right
C. go farther to the center
D. not be affected

19. The term *Thiele* refers to a type of 19.____
A. balance B. thermometer
C. tube D. viscosimeter

20. The term *Abbé* refers to a type of apparatus generally used to determine 20.____
A. boiling point elevation B. density
C. refractive index D. surface tension

21. The pycnometer is GENERALLY used to determine 21.___
    A. density B. mass
    C. melting point D. viscosity

22. The cryoscope is GENERALLY used to determine the 22.___
    A. boiling point B. density
    C. freezing point D. surface tension

23. Of the following carbohydrates, the one which is NOT classified as a polysaccharide is 23.___
    A. cellulose B. inulin C. invertose D. starch

24. Aliphatic means, *most nearly*, 24.___
    A. acidic B. cyclic
    C. nitrogen-containing D. non-cyclic

25. Ethyl alcohol is often made UNSUITABLE for drinking by the addition of 25.___
    A. copper sulphate B. methyl alcohol
    C. sodium chloride D. water

KEY (CORRECT ANSWERS)

1. B
2. B
3. D
4. A
5. B

6. D
7. A
8. A
9. A
10. C

11. A
12. D
13. A
14. B
15. D

16. A
17. C
18. A
19. C
20. C

21. A
22. C
23. C
24. D
25. B

TEST 2

DIRECTIONS: Each question or incomplete statement is followed by several suggested answers or completions. Select the one that BEST answers the question or completes the statement. *PRINT THE LETTER OF THE CORRECT ANSWER IN THE SPACE AT THE RIGHT.*

1. The industrial process of transforming vegetable oils into solid fats is an example of 1.___
   A. halogenation B. homogenization
   C. hydrogenation D. hydrolysis

2. Substances which are INTERMEDIATE between proteins and amino acids are called 2.___
   A. albumins B. caseins C. gelatines D. peptones

3. The enzyme which hydrolyzes proteins into amino acids is 3.___
   A. amylase B. emulsin C. trypsin D. saccharase

4. Hydrolysis of an ester in basic solution is GENERALLY termed 4.___
   A. fermentation B. halogenation
   C. hydrogenation D. saponification

5. TEMPORARY hardness of water is generally removed by the addition of 5.___
   A. bicarbonates B. carbonates
   C. chlorides D. hydroxides

6. PERMANENT hardness of water is generally removed by 6.___
   A. aeration B. decantation
   C. evaporation D. precipitation

Questions 7-16.

DIRECTIONS: Column I lists terms, each of which can be properly matched with the laboratory apparatus in Column II. For each item of Column I, write in the answer space the letter in front of the option in Column II with which it is generally associated.

Column I	Column II	
7. Arnold	A. balance	7.___
8. Beckmann	B. beaker	8.___
9. Duboscq	C. bottle	9.___
10. Jackson	D. burner	10.___
11. Kipp	E. candle turbidimeter	11.___

Column I	Column II	
12. Meker	F. colorimeter	12.___
13. Nessler	G. funnel	13.___
14. Ostwals	H. gas generator	14.___
15. Westphal	J. steam sterilizer	15.___
16. Woulff	K. thermometer	16.___
	L. tube	
	M. viscosimeter	

Questions 17-25.

DIRECTIONS: Column I lists the names of metallic ions, each of which can be placed with one of the analytical groups in Column II. For each item of Column I, write in the answer space the letter in front of the group in Column II to which it belongs. (Where ions belong to more than one group, so indicate.)

17. Aluminum	A. Hydrochloric Acid Group	17.___
18. Ammonium	B. Hydrogen Sulfide Group	18.___
19. Bismuth	C. Ammonium Sulfide Group	19.___
20. Iron	D. Ammonium Carbonate Group	20.___
21. Lead	E. Soluble Group	21.___
22. Mercury		22.___
23. Strontium		23.___
24. Tin		24.___
25. Zinc		25.___

KEY (CORRECT ANSWERS)

1. C	6. D	11. H	16. C	21. A,B
2. D	7. J	12. D	17. C	22. A,B
3. C	8. K	13. L	18. E	23. D
4. D	9. F	14. M	19. B	24. B
5. D	10. E	15. A	20. C	25. C

EXAMINATION SECTION

TEST 1

DIRECTIONS: Each question or incomplete statement is followed by several suggested answers or completions. Select the one that BEST answers the question or completes the the statement. *PRINT THE LETTER OF THE CORRECT ANSWER IN THE SPACE AT THE RIGHT.*

1. The HOTTEST portion in a flame from a bunsen burner is located _____ of the flame. 1.___
   A. at the top of the outer yellow cone
   B. at the top of the inner blue cone
   C. at the middle of the inner blue cone
   D. about 1 cm above the very top

2. In which one of the following ways should a piece of small glass tubing be cut? 2.___
   A. Hold the tubing in a flame and pull the two ends apart
   B. Scratch the tubing with a file and bend the tube toward the scratch
   C. Scratch the tubing with a file and bend the tubing away from the scratch
   D. Heat the tubing in a flame and allow a drop of cold water to fall where you want the cut

3. Which one of the following materials is *transparent* in infrared light? 3.___
   A. NaCl crystals B. Glass
   C. Quartz D. Pyrex

4. Which one of the following solids will *dissolve* in concentrated aqueous NH_3? 4.___
   A. $Al(OH)_3$ B. $Fe(OH)_3$ C. $Cu(OH)_2$ D. $Pb(OH)_2$

5. The one of the following ions which will cause a yellow color in a flame test is 5.___
   A. K^+ B. Cu^{++} C. Na^+ D. Ca^{++}

6. Which one of the following ions can NOT be precipitated in the presence of excess NaOH? 6.___
   A. Fe^{+++} B. Ni^{+++} C. Co^{+++} D. Al^{+++}

7. The one of the following poisonous gases which has no odor is 7.___
   A. CO B. HCN C. NH_3 D. H_2S

8. Which one of the following indicators is *colorless* in acid and *red* in very basic solutions? 8.___
   A. Litmus B. Methyl violet
   C. Bromethymol blue D. Phenolphthalein

9. The one of the following compounds which is known as vitamin C is 9.___
   A. cyanocobalamin B. ascorbic acid
   C. boric acid D. arginine

10. The reaction between a conjugated diene and an alkene to form a six-membered ring compound is called 10.___
   A. esterification B. a Diels-Adler reaction
   C. an Aldol condensation D. a Markownikoff addition

11. The chemical name for aspirin is 11.___
   A. methylanthranilate B. acetylsalicyclic acid
   C. phenacetin D. methylsalicylate

12. Which one of the following compounds is NOT an organic base? 12.___
   A. Aniline B. Pyridine
   C. Trimethyamine D. Tetramethyammonium iodide

13. Which one of the following instruments is SUITABLE for the determination of ethanol in benzene? 13.___
A(n)
   A. pH meter
   B. gas chromatograph
   C. atomic absorption spectrometer
   D. visible spectrometer

14. How many colloidal suspensions often be caused to coagulate and precipitate? 14.___
By
   A. treatment with cold water
   B. occulusion
   C. addition of NH_4Cl
   D. centrifugation

15. The addition of $CaCl_2$ to water causes this compound to become 15.___
   A. hard B. acidic C. soft D. basic

16. The Mohr and Volhard methods are used to determine the presence of 16.___
   A. acid B. chloride
   C. density D. molecular weight

17. The mean deviation of the three values of 2, 6, and 7 is 17.___
   A. 0 B. 2 C. 5 D. 6

18. A cation exchange resin can be used to remove _____ from water. 18.___
   A. Cl^- B. OH^- C. Na^+ D. alcohol

19. Which one of the following groups contains solvents which are all separately miscible with water at room temperature? 19.___
   A. Acetone, benzene, sulfuric acid
   B. Acetone, methanol, acetic acid
   C. Ether, methanol, ethanol
   D. Chloroform, carbontetrachloride, petroleum ether

20. The absorption of organic compounds on alumina followed by separation by elution with different solvents is called 20.___
   A. filtration
   B. distillation
   C. crystallization
   D. chromatography

KEY (CORRECT ANSWERS)

1. B	11. B
2. C	12. D
3. A	13. B
4. C	14. C
5. C	15. A
6. D	16. B
7. A	17. B
8. D	18. C
9. B	19. B
10. B	20. D

TEST 2

DIRECTIONS: Each question or incomplete statement is followed by several suggested answers or completions. Select the one that BEST answers the question or completes the statement. *PRINT THE LETTER OF THE CORRECT ANSWER IN THE SPACE AT THE RIGHT.*

1. The Fisher reagent (I_2 + SO_2 + C_5H_5N in CH_3OH) is used for 1.___
   A. the identification of amines
   B. the determination of the presence of H_2O in organic solvents
   C. the separation of amino acids
   D. testing for chloride ion

2. Which one of the following is an analytical method for the determination of the presence of nitrogen in organic compounds? 2.___
   A. The Kjeldahl method
   B. Treatment by the Versene (EDTA) method
   C. Iodometry
   D. X-ray diffraction method

3. Racemization of d-Tartaric acid will produce _____ tartaric acid. 3.___
   A. meso- B. l- C. d,l- D. d-

4. Which one of the following reagents is used for the detection of amino acids by thin layer chromatography? 4.___
   A. Aniline phthalate B. Bromothymol blue
   C. Benzene D. Ninhydrin

5. What is the molar concentration of a material in solution in a 1 cm cell, if the optical density is 1.0 at a wave length where the solvent is transparent and the solute has an extinction coefficient of 10 liters/mole cm? 5.___
   A. 0.1M B. 1.0M C. 10M D. 100M

6. What is the measurement of a standard taper 24/40 stopper? _____ mm at the top and _____ mm _____. 6.___
   A. 24; 40; at the bottom B. 40; 24; at the bottom
   C. 24; 40; long D. 40; 24; long

7. To what extent should a mercury thermometer calibrated for total immersion be immersed in a test solution? 7.___
   A. Totally
   B. Only 76 mm
   C. Sufficiently to reach the mercury column
   D. Only to totally cover the bulb of mercury at the bottom of the thermometer

8. The white precipitate formed when sodium thiosulfate ($Na_2S_2O_3$) solution is treated with acid is 8.___
   A. $H_2S_2O_3$ B. SO_3 C. S D. Na_2SO_4

9. The one of the following atoms which possesses the GREATEST electronegativity is 9.___
   A. Na B. C C. O D. Al

10. The oxidation states of cobalt in water solutions are 10.___
    A. +1 and +2 B. +2 and +3
    C. +3 and +4 D. +2 and +4

11. Which one of the following is an example of a reactive hydride? 11.___
    A. CH_4 B. SiH_4 C. NH_3 D. NaH

12. The infrared absorption indicative of an organic ketone is found at about _____ cm^{-1}. 12.___
    A. 1000 B. 1700 C. 3000 D. 10000

13. Which one of the following compounds does NOT show any NMR spectrum at 60 megacycles? 13.___
    A. CH_3OH B. CCl_4 C. C_6H_6 D. SiH_4

14. At which one of the following temperatures is the density of water GREATEST? 14.___
    A. -4°C B. 0°C C. 4°C D. 100°C

15. Which one of the following forms of mercury offers the GREATEST environmental threat? 15.___
    A. Hg metal B. CH_3HgCH_3 C. $HgCl_2$ D. Hg_2Cl_2

16. What is the APPROXIMATE melting point for a mixture of 99% benzoic acid (mp 122°C) and 1% succinic acid (mp 185°C)? 16.___
    A. Below 122°C B. Between 122° and 185°C
    C. Above 130°C D. 185°

17. The number of grams in 1 pound (avdp) is 17.___
    A. 3.3 B. 62.5 C. 100 D. 453

18. A water solution contains either Na_2SO_4 or Na_2CO_3 and nothing else. 18.___
    To determine which salt is present, you should
    A. use a flame test
    B. make the solution acid with HCl and add a $BaCl_2$ solution
    C. make the solution basic with NaOH and add a $BaCl_2$ solution
    D. add a $BaCl_2$ solution only

19. Ignition of ferric hydroxide in air produces 19.___
    A. iron metal B. FeO
    C. Fe_2O_3 D. Fe_3O_4

20. Coulometry is a technique which involves measurement of 20.___
A. density
B. gas volume
C. current flow
D. color shading

KEY (CORRECT ANSWERS)

1. B	11. D
2. A	12. B
3. C	13. B
4. D	14. C
5. A	15. B
6. C	16. A
7. A	17. D
8. C	18. B
9. C	19. C
10. B	20. C

EXAMINATION SECTION

TEST 1

DIRECTIONS: Each question or incomplete statement is followed by several suggested answers or completions. Select the one that BEST answers the question or completes the statement. *PRINT THE LETTER OF THE CORRECT ANSWER IN THE SPACE AT THE RIGHT.*

1. The valence of an element tells 1.___
   A. its atomic weight
   B. the solubility of its compounds
   C. its stability
   D. how many electrons its atom lends, borrows, or shares
   E. how many compounds can be formed

2. One way in which compounds differ from mixtures is that compounds have 2.___
   A. more elements
   B. fewer elements
   C. greater activity
   D. a definite composition
   E. physical properties

3. Sugar is a compound because it 3.___
   A. is granular
   B. gives off water when heated
   C. blackens when heated
   D. has a definite composition
   E. is sweet

4. Argon and neon are inert elements because their atoms have 4.___
   A. complete outermost rings
   B. no electrons
   C. chemical reactivity
   D. no protons
   E. no neutrons

5. Any solution which conducts an electric current is called a(n) 5.___
   A. ion
   B. electrolyte
   C. non-electrolyte
   D. electrode
   E. catalyst

6. Hydrogen is contained in all 6.___
   A. liquids
   B. oxides
   C. salts
   D. compounds
   E. acids

7. Plants give off oxygen during the day because 7.___
   A. the body of the plant needs to breathe oxygen to live
   B. oxygen is one of the products of photosynthesis
   C. oxygen is one of the products of respiration
   D. burning of food produces energy
   E. the extra warmth acts as a catalytic agent

8. A manufacturing city is relatively free from smoke. This is often due to 8.___
   A. the use of acids in manufacturing
   B. an inventor who has found a new use of by-products

C. the process of collecting carbon by charged electrodes
D. the exclusive use of steam boilers in all manufacturing establishments
E. the fact that the city is in a region of high-pressure areas

9. The substance used in etching glass is 9.____
A. sulfuric acid B. tincture of iodine
C. hydrobromic acid D. aqua regia
E. hydrofluoric acid

10. Burning of wood is an example of 10.____
A. an endothermic reaction B. reduction
C. chemical change D. physical change
E. catalysis

11. A gas collected in a bottle was allowed to mix with a small amount of air. A burning splint applied to the mouth of the bottle caused a sharp explosion. 11.____
The gas collected was
A. oxygen B. chlorine C. hydrogen
D. nitrogen E. sulfur dioxide

12. Diamond is a(n) 12.____
A. alloy B. metal
C. crystal D. ore
E. inorganic compound

13. If water is saturated with salt (sodium chloride), the freezing point is 13.____
A. the same as for distilled water
B. lower than for pure water
C. higher than for pure water
D. zero degrees Centigrade
E. plus four degrees Centigrade

14. If a relatively small amount of solute is dissolved in a relatively large amount of solvent, the solution is called 14.____
A. dilute B. saturated C. ionic
D. concentrated E. none of the above

15. In a Bunsen burner, luminous flames result when the 15.____
A. gas is turned too low
B. gas carbon dioxide is formed
C. barrel of the burner is too long
D. spud is closed
E. air openings are closed

16. Sodium hydroxide is a(n) 16.____
A. acid B. salt C. oxide
D. base E. carbohydrate

17. Which of the following molecules contains five atoms? 17.____
A. $Ca_3(PO_4)_2$ B. Na_2CO_3 C. Al_2O_3
D. $CaSO_4$ E. NH_4OH

18. Neutralization reactions usually form water and 18.___
A. acids B. bases C. oxides
D. salts E. absorbing agents

19. Why do chemical reactions of the body take place so much faster within the body than they do outside the body? Because 19.___
A. the body is active
B. the reactions are those of the body and designed to go on in the body
C. the reactions of the body are organic reactions
D. there are enzymes in the body which speed up reactions
E. the blood is a good solvent

20. The study of carbon compounds is known as ____ chemistry. 20.___
A. inorganic B. organic C. qualitative
D. analytical E. physical

21. Sodium compounds, when heated in a flame, produce the color 21.___
A. red B. yellow C. green D. violet D. pink

22. Substances which dissociate into ions when in solution are called 22.___
A. electrolytes B. non-electrolytes C. hydrates
D. aqueous E. catalysts

23. A hydrocarbon with the formula CH_4 is 23.___
A. methylene B. acetylene C. ethane
D. propane E. methane

24. Cotton is MOST often bleached by use of 24.___
A. ozone B. hydrogen peroxide
C. chlorine D. sulfur dioxide
E. hydrochloric acid

25. The formula for milk of magnesia is 25.___
A. $Mg(OH)_2$ B. MgO C. $MgSO_4$
D. $MgCl_2$ E. $Mg_3(PO_4)_2$

26. Natural gas is composed MAINLY of 26.___
A. methane B. carbon monoxide C. hydrogen
D. acetylene E. carbon dioxide

27. What gas in the atmosphere is necessary for the operation of an automobile engine? 27.___
A. Carbon dioxide B. Hydrogen C. Helium
D. Carbon monoxide E. Oxygen

28. An element has an atomic weight of 20 and its atomic number is 10. 28.___
The number of planetary electrons about the nucleus of the element is
A. 0 B. 10 C. 20 D. 30 E. 40

29. Concentrated sulfuric acid is used in the preparation of hydrogen chloride because it 29.____
A. reacts slowly
B. is very concentrated
C. is a heavy acid
D. is a dehydrating agent
E. has a high boiling point

30. A hydrocarbon with the formula C_2H_2 is 30.____
A. methyl chloride
B. methylene
C. chloroform
D. iodoform
E. acetylene

31. When iron rusts, it 31.____
A. gains in weight
B. remains chemically unchanged
C. loses weight
D. combines with hydrogen
E. is reduced

32. Equations are balanced by 32.____
A. the use of exponents
B. the use of subscripts
C. the use of coefficients
D. placing parentheses around radicals
E. placing all products on the right side of the equation

33. Chlorine is obtained commercially CHIEFLY by 33.____
A. action of acids on metals
B. heating sodium chloride
C. electrolysis of hydrochloric acid
D. electrolysis of brine
E. heating chlorides

34. Slow oxidation differs from burning in that 34.____
A. no heat is given off
B. more time is required
C. oxides are formed
D. dioxides are formed
E. compounds are formed

35. Which one of the following does NOT produce a high degree of ionization when in solution? 35.____
A. HCl
B. H_2SO_4
C. NaOH
D. KCl
E. CH_3OH

36. The anhydride of H_2SO_3 is 36.____
A. SO_3
B. SO_2
C. SO_4
D. H_2O
E. S_2O_3

37. Acids whose names end in -ous have salts with names ending in 37.____
A. -ide
B. -ate
C. -ite
D. -ine
E. -ous

38. The general formula for organic acids is 38.____
A. RCOOH
B. ROH
C. RCHO
D. ROR
E. RCOOR

39. Which one of the following iron ores contains the GREATEST percentage of iron? (Atomic weights: Fe = 56, O = 16, C = 12, S = 32, H = 1.) 39.____
A. $Fe_2O_3(H_2O)_3$
B. FeS_2
C. $FeCO_3$
D. Fe_2O_3
E. Fe_3O_4

40. Which is the CORRECT formula for magnesium nitride? 40.____
A. MgN
B. Mg_3N_2
C. $Mg(NO_2)_2$
D. $Mg(NO_3)_2$
E. Mg_3N

KEY (CORRECT ANSWERS)

1. D	11. C	21. B	31. A
2. D	12. C	22. A	32. C
3. D	13. B	23. E	33. D
4. A	14. A	24. C	34. B
5. B	15. E	25. A	35. E
6. E	16. D	26. A	36. B
7. B	17. C	27. E	37. C
8. C	18. D	28. B	38. A
9. E	19. D	29. E	39. E
10. C	20. B	30. E	40. B

TEST 2

DIRECTIONS: Each question or incomplete statement is followed by several suggested answers or completions. Select the one that BEST answers the question or completes the statement. *PRINT THE LETTER OF THE CORRECT ANSWER IN THE SPACE AT THE RIGHT.*

Questions 1-14.

DIRECTIONS: Questions 1 through 14 contain 7 sets of 2 questions each. The first question in each set is an information question, and the second is a question on the chemical principle that best explains the answer to the preceding question. In each case, put the letter of your answer in the space at the right.

1. A storekeeper placed some calcium chloride between his sash and storm window during a severe cold spell. What happened? 1.___
   A. The windows became more heavily frosted than the nearby windows.
   B. The substance became dry and powder-like.
   C. The calcium chloride evaporated.
   D. The windows remained almost free of frost.
   E. Nothing happened.

2. Which one of the following statements gives the principle that BEST explains the answer to Question 1? Many 2.___
   A. gases such as water vapor and carbon dioxide can be solidified
   B. chemical reactions depend upon the lack of or presence of moisture
   C. solids such as dry ice and iodine pass directly from a solid to a gaseous state
   D. substances readily lose their water of hydration or crystallization on exposure to air
   E. substances have the property of absorbing or adsorbing moisture from air or other substances

3. If a positive and a negative electrode are introduced into a solution of copper chloride, the copper will be deposited on the negative electrode. This indicates 3.___
   A. that the copper particles in the solution had a positive charge
   B. that the copper particles had acquired a negative charge from the negative electrode
   C. that the copper particles were electrically neutral
   D. that the chloride particles were repelled by the copper particles
   E. nothing at all

4. Which one of the following statements gives the principle that BEST explains the answer to Question 3? 4.___
   A. Colloidal suspensions of different metals impart different colors to a solution.
   B. When salts ionize, the metal gives up electrons to the non-metal.
   C. In combining with other elements, metals gain electrons.
   D. Like charges repel each other and unlike charges attract.
   E. Basic metals are protected from corrosion by plating them with relatively non-corrosive metals.

5. Sodium chloride and sulfuric acid were placed in a flask and heated gently. The hydrochloric acid gas was passed over water and dissolved, thereby making a solution of hydrochloric acid. 5.___
Why is HCl the only product given off as a gas?
   A. The heat produces a chemical reaction.
   B. The sodium chloride furnishes the chloride ion.
   C. The acid furnishes the hydrogen ion necessary to make an acid.
   D. HCl is the only product volatile at the temperature of the reaction.
   E. The sodium sulfate is a precipitate.

6. Which one of the following statements gives the principle that BEST explains the answer to Question 5? 6.___
   A. Compounds possessing different boiling points may be separated by destructive distillation.
   B. Acids are produced only by double decomposition reactions.
   C. All chemical reactions are speeded up by heat.
   D. Acids furnish the hydrogen ion in varying degrees of concentration.
   E. Acids with low boiling points may be prepared by using an acid with a high boiling point.

7. A boy placed some zinc in a flask with some dilute hydrochloric acid. 7.___
What happened?
   A. Hydrogen was given off.
   B. Chlorine was given off.
   C. Oxygen was given off.
   D. Zinc chloride was precipitated.
   E. Nothing happened.

8. Which one of the following statements gives the principle that BEST explains the answer to Question 7? 8.___
   A. Metals replace hydrogen from all acids.
   B. Gases are produced by simple replacement reactions.
   C. Elements are able to replace those below them in the electrochemical series.
   D. Chemical opposites, such as a metal and an acid, usually unite.
   E. Zinc is a very inactive metal.

9. Hydrogen gas was passed into a test tube containing copper oxide. Heat was then applied to the tube. 9.____
What probably happened?
A. An explosion occurred.
B. Water was formed in the tube.
C. A reddish-colored gas was given off.
D. A thin green coating appeared on the copper.
E. Nothing happened.

10. Which one of the following statements gives the principle that BEST explains the answer to Question 9? 10.____
A. Copper is a comparatively inactive metal.
B. Metal oxides often combine with an acid to form a carbonate.
C. Many oxygen compounds when heated with hydrogen will release oxygen.
D. Noticeable heat and light occur in many chemical reactions.
E. Some chemical compounds are able to absorb water from air.

11. One series of hydrocarbons contains these compounds: CH_4, C_2H_6, C_3H_8. 11.____
What is the general formula for finding the formulas of additional members of this series?
A. C_nH_{2n} B. C_nH_{2n+2} C. C_nH_{2n+4}
D. C_nH_{2n-2} E. C_nH_{2n+6}

12. Which one of the following statements gives the principle that BEST explains the answer to Question 11? 12.____
A. Unsaturated hydrocarbons are capable of taking on additional hydrogen.
B. Saturated hydrocarbons are incapable of holding more hydrogen.
C. Organic compounds are compounds in which the elements share electrons.
D. Saturated hydrocarbons form what is known as substitution products.
E. In a homologous series, each member differs from the next by the same group of elements.

13. A bottle was filled almost to the top with sulfuric acid and left unstoppered. 13.____
What MOST likely happened?
A. The amount of sulfuric acid was increased.
B. The sulfuric acid became more concentrated.
C. The bottle overflowed.
D. Sulfur dioxide was given off.
E. Nothing happened.

14. Which one of the following statements gives the principle that BEST explains the answer to Question 13? 14.___
    A. Evaporation takes place at the surface of the liquid.
    B. Unstable compounds often give off gases in decomposing.
    C. Dissociation or ionization has made possible the production of new elements.
    D. Most acids will expand in volume if exposed to the air.
    E. Some chemical compounds are able to absorb water from the air.

Questions 15-20.

DIRECTIONS: Each of Questions 15 through 20 consists of five terms, one of which does NOT belong with the other four. You are to select the term which does NOT belong. Mark its letter in the space at the right.

15. A. nitrogen B. magnesium C. tin D. antimony E. bismuth 15.___

16. A. boiling point B. inert C. solubility D. specific gravity E. density 16.___

17. A. aldehyde B. ketone C. nitride D. ester E. ether 17.___

18. A. dissolving B. evaporating C. freezing D. burning E. solidifying 18.___

19. A. Bunsen B. Cavendish C. Einstein D. Frasch E. Priestley 19.___

20. A. $CaCO_3$ B. $CaSO_4$ C. $Ca_3(PO_4)_2$ D. $CaCl_2$ E. $CaSiO_3$ 20.___

Questions 21-24.

DIRECTIONS: Questions 21 through 24 are based on the following data:

RESULTS OF THREE EXPERIMENTS TO DETERMINE WHAT FACTORS, IF ANY, AFFECT THE SOLUBILITY OF SOLIDS

Experiment	Kind and Amount Of Solvent	Temperature of Solvent	Pressure	Kind of Solute	Amount of Solute Dissolved
I	100 cc CCl_4	10°C.	600 mm	Sulfur	15 grams
	100 cc CCl_4	10°C.	700 mm	Sulfur	15 grams
II	100 cc H_2O	100°C.	700 mm	Sulfur	15 grams
	100 cc H_2O	50°C.	800 mm	Sulfur	1 gram
III	100 cc H_2O	10°C.	700 mm	Sulfur	.1 gram
	100 cc CCl_4	10°C.	700 mm	Sulfur	15 grams

Constant factors are those which are kept the same in an experiment. Varied factors are those which are not kept the same. You are to assume that if a factor was found NOT to affect the solubility of solids in any one experiment, it would not do so in any other experiment. The number of grams of solute dissolved in the given volume of solvent produced saturation.

Read each question and study the table carefully. Decide which answer is best and mark its letter in the space at the right.

21. What factor or combination of factors was constant in Experiment I? 21.____
 A. Temperature of solvent
 B. Kind of solute; pressure
 C. Temperature of solvent; pressure; kind of solute
 D. Kind of solvent; pressure; kind of solute
 E. Kind of solvent; temperature of solvent; kind of solute

22. What factor or combination of factors was varied in Experiment II? 22.____
 A. Temperature of solvent
 B. Kind of solute; temperature of solvent
 C. Pressure; temperature of solvent
 D. Kind of solvent; pressure
 E. Kind of solute; kind of solvent

23. What factor in Experiment III produced the observed difference in the amount of solute dissolved? 23.____
 A. Kind of solvent
 B. Temperature of solvent
 C. Pressure
 D. Kind of solute
 E. None of the above

24. In these experiments, what factors definitely affect the solubility of solids? 24.____
 A. Kind of solute; pressure
 B. Kind of solvent; kind of solute
 C. Temperature of solvent; pressure
 D. Pressure; temperature of solvent; kind of solvent
 E. Temperature of solvent; kind of solvent

Questions 25-29.

DIRECTIONS: Questions 25 through 29 are based on the figure below. Study the figure and read each question carefully. Then, in the space at the right, mark the letter of the BEST answer.

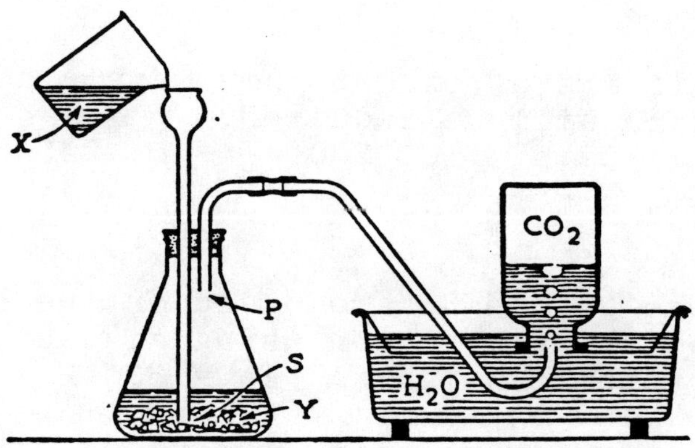

25. If this reaction is to work successfully, Beaker X should contain 25.___
    A. sodium hydroxide B. copper sulfate
    C. hydrochloric acid D. carbon tetrachloride
    E. sodium chloride

26. If this reaction is to work successfully, Flask Y should contain 26.___
    A. calcium sulfate B. calcium hydroxide
    C. zinc sulfate D. calcium carbonate
    E. calcium oxide

27. The MOST common chemical test for the presence of carbon dioxide makes use of 27.___
    A. sodium hydroxide B. hydrochloric acid
    C. ammonium hydroxide D. calcium sulfate
    E. calcium hydroxide

28. The ends of the glass tubing, S and P, 28.___
    A. are in correct position
    B. should be reversed in their respective positions
    C. should both be just above the liquid
    D. should both be near the top of the flask
    E. should both be in the liquid

29. Which of the following substances should NOT be used at Y? 29.___
    A. Limestone B. Coral C. Gypsum
    D. Chalk E. Marble

Questions 30-31.

DIRECTIONS: Questions 30 and 31 are based on the equation for the preparation of oxygen by heating potassium chlorate. (Atomic weights: K = 39, Cl = 35.5, O = 16.)

30. To balance the equation, the coefficient of the potassium chlorate is 30.____
A. 1 B. 2 C. 3 D. 4 E. 5

31. To obtain 67.2 liters of oxygen, how many grams of potassium chlorate will be required? 31.____
A. 74 B. 122.5 C. 200 D. 245 E. 367.5

Questions 32-33.

DIRECTIONS: Questions 32 and 33 are based on the equation for the preparation of hydrogen using zinc and hydrochloric acid. (Atomic weights: Zn = 65, H = 1, Cl = 35.5.)

32. What is the equation or reacting weight of zinc chloride? 32.____
A. 65 B. 100.5 C. 125 D. 136 E. 165.5

33. If 65 grams of zinc are used, how many grams of hydrogen will be formed? 33.____
A. 1 B. 2 C. 4 D. 6 E. 8

34. What will be one of the ions in the ionic equation for the formation of $CuSO_4$? 34.____
A. Cu^+ B. Cu^{++} C. SO_3^{--} D. SO_4^- E. SO_4^{---}

35. Which is the CORRECT formula for aluminum oxide? 35.____
A. Al_2O_3 B. AlO C. Al_3O_2 D. Al_3O_4 E. Al_2O_5

36. How many electrons will the iron lose when $FeCl_2$ is changed to $FeCl_3$? 36.____
A. 1 B. 2 C. 3 D. 4 E. 6

37. Which is the CORRECT formula for silver bromide? 37.____
A. AgBr B. Ag_2Br C. Ag_2BrO_3 D. $Ag_2(Br)_3$ E. Ag_3Br_2

38. What is one of the products in the completely balanced equation $Na_2SO_3 + H_2SO_4$ (dilute)→ ? 38.____
A. Na_2SO_3 B. SO_3 C. Na_2SO_4 D. $NaHSO_3$ E. H_2S

39. The electrolysis of water represents which one of the following chemical changes? 39.____
A. Synthesis
B. Simple decomposition
C. Simple replacement
D. Double replacement
E. Hydrolysis

40. Which is the CORRECT formula for ammonium sulfate? 40.____
A. NH_4SO_4 B. NH_4SO_3 C. NH_4S D. $(NH_4)_2SO_4$ E. $(NH_4)_2SO_3$

KEY (CORRECT ANSWERS)

1. D	11. B	21. E	31. D
2. E	12. E	22. C	32. D
3. A	13. C	23. A	33. B
4. D	14. E	24. E	34. B
5. D	15. A	25. C	35. A
6. E	16. B	26. D	36. A
7. A	17. C	27. E	37. A
8. C	18. D	28. A	38. C
9. B	19. C	29. C	39. B
10. C	20. D	30. B	40. D

EXAMINATION SECTION

TEST 1

DIRECTIONS: Each question or incomplete statement is followed by several suggested answers or completions. Select the one that BEST answers the question or completes the statement. *PRINT THE LETTER OF THE CORRECT ANSWER IN THE SPACE AT THE RIGHT.*

Questions 1-4.

DIRECTIONS: Questions 1 through 4 are based upon the following equation representing the principal method of producing methyl alcohol: $CO + 2H_2 = CH_3OH$. As the alcohol is a gas under conditions of the reaction, the reaction takes place entirely in the gaseous state.

1. The presence of a catalyst will 1.___
   A. *increase* the yield of methyl alcohol
   B. *increase* the rate of the reaction and increase the yield of methyl alcohol
   C. *decrease* the time required to reach equilibrium but have no effect on the equilibrium concentrations
   D. *decrease* the time required to reach equilibrium and decrease the yield of methyl alcohol
   E. *increase* the amount of heat evolved during the reaction

2. As the temperature is raised, the yield of CH_3OH is decreased. 2.___
   This indicates
   A. an increase in temperature decreases the rate of a reaction
   B. the reaction as written absorbs heat
   C. the reaction as written evolves heat
   D. an increase in temperature increases the rate of reaction
   E. the catalyst is heat sensitive

3. If after the equilibrium is established, the pressure is increased, the yield of CH_3OH will 3.___
   A. be increased
   B. be decreased
   C. not be affected
   D. depend on the catalyst being used
   E. be 32 grams

4. For every 1000 cubic feet of hydrogen that react, 4.___
   A. 1000 cubic feet of carbon monoxide react
   B. 1000 cubic feet of methyl alcohol are formed
   C. 2000 cubic feet of methyl alcohol are formed
   D. 500 cubic feet of methyl alcohol are formed
   E. 1000 gallons of methyl alcohol are formed

Questions 5-13.

DIRECTIONS: Answer Questions 5 through 13 based upon the following atom. An assumed atom has the following structure: The nucleus contains 46 neutrons and 35 protons. The first three electron orbits contain 2, 8, 18 electrons.

5. The atomic weight is 5.____
A. 28 B. 35 C. 46 D. 70 E. 81

6. The atomic number is 6.____
A. 28 B. 35 C. 46 D. 70 E. 81

7. The negative oxidation number of the element would *probably* be 7.____
A. -1 B. -2 C. -3 D. -4 E. -7

8. The MAXIMUM POSITIVE oxidation number of the element would *probably* be 8.____
A. 1 B. 3 C. 5 D. 7 E. 8

9. In concentrated sulfuric acid, the bond between hydrogen and oxygen is BEST classified as a(n) _____ bond. 9.____
A. ionic
B. covalent
C. hydrogen
D. coordinate covalent
E. ion-dipole

10. In sulfuric acid, the bond between sulfur and the oxygen to which no hydrogen is attached is a(n) _____ bond. 10.____
A. ionic
B. coordinate covalent
C. double
D. ion dipole
E. nonpolar covalent

11. A 1 normal solution of sulfuric acid, molecular weight 98, contains _____ grams solute per liter. 11.____
A. 1 B. 2 C. 49 D. 98 E. 196

12. _____ will conduct an electric current. 12.____
A. Carbon tetrachloride
B. Hydrogen chloride dissolved in toluene
C. Hydrogen chloride dissolved in water
D. Sugar dissolved in water
E. Glacial acetic acid

13. Solid iodine is heated, converted to a purple gas, and resolidified on the surface of the container. This process is known as 13.____
A. precipitation
B. condensation
C. evaporation
D. sublimation
E. purification

Questions 14-19.

DIRECTIONS: Answer Questions 14 through 19 concerning the following reaction: $Cu + 2H_2SO_4 \rightarrow CuSO_4 + 2H_2O + SO_2$ using the atomic weights Cu = 63.5, H = 1, S = 32, O = 16.

14. The substance reduced is 14.___
A. hydrogen
B. copper
C. sulfur
D. oxygen
E. water

15. The reducing agent is 15.___
A. copper
B. sulfuric acid
C. cupric sulfate
D. water
E. sulfur dioxide

16. If 31.75 grams of copper are used, the volume of SO_2 gas produced at STP is _____ liters. 16.___
A. 2
B. 11.2
C. 22.4
D. 32
E. 64

17. The substance which loses electrons is 17.___
A. copper
B. hydrogen
C. sulfur
D. oxygen
E. water

18. If an excess H_2SO_4 is present, the amount of copper needed to produce 3 gram moles of $CuSO_4$ is 18.___
A. 3
B. 21.1
C. 63.5
D. 254
E. correct answer not given

19. _____ gram(s) of water are produced when 9.8 grams H_2SO_4 react with 7 grams of Cu. 19.___
A. 1.0
B. 1.8
C. 2
D. 3.6
E. 4

20. The oxidation number of sulfur in Ag_2S is 20.___
A. 2
B. 1
C. 0
D. -1
E. -2

21. The oxidation number of oxygen in Al_2O_3 is 21.___
A. +3
B. +2
C. 0
D. -2
E. -6

22. For the reaction $4HCl(g) + O_2(g) = 2H_2O(g) + 2Cl_2(g) +$ 27 Kcal, the equilibrium concentration of Cl_2 would be increased by 22.___
A. increasing the temperature
B. adding a catalyst
C. adding water
D. increasing the pressure
E. all the preceding procedures would increase the equilibrium concentration of Cl_2

23. For the reaction $2SO_2(g) + O_2(g) = 2SO_3(g) + 23$ Kcal, the equilibrium concentration of SO_2 would be increased by 23.___
A. an increase in temperature
B. an increase in pressure
C. removing SO_3
D. addition of oxygen
E. addition of catalyst

24. A substance that exists *mainly* as molecules when dissolved in water is 24.___
A. NaCl B. NaOH C. CH_3OH D. HCl E. $CaBr_2$

25. The solubility product for $PbCl_2$ is equal to 25.___
A. $\{Pb^{++}\} + 2\{Cl^-\}$
B. $\{Pb^{++}\} + \{Cl^-\}$
C. $\{Pb^{++}\}\{Cl\}$
D. $\{Pb^{++}\}\{Cl\}^2$
E. $\{Pb^{++}\}\ 2\ \{Cl^-\}^2$

KEY (CORRECT ANSWERS)

1. C
2. C
3. A
4. D
5. E

6. B
7. A
8. D
9. B
10. B

11. C
12. C
13. D
14. C
15. A

16. B
17. A
18. E
19. B
20. E

21. D
22. D
23. A
24. C
25. D

TEST 2

DIRECTIONS: Each question or incomplete statement is followed by several suggested answers or completions. Select the one that BEST answers the question or completes the statement. *PRINT THE LETTER OF THE CORRECT ANSWER IN THE SPACE AT THE RIGHT.*

1. The oxidation number of chromium in $K_2Cr_2O_7$ is 1.___
   A. -2 B. 0 C. 2 D. 6 E. 12

2. A metal that will react with acids to produce hydrogen gas is 2.___
   A. Ag B. Cu
   C. Hg D. Al
   E. all of listed metals

3. A metal that will react with cold water to produce hydrogen gas is 3.___
   A. Al B. Zn C. Cu D. K E. Hg

4. In general, nonmetals 4.___
   A. are electric conductors
   B. displace hydrogen from dilute acid solutions
   C. exist as positive ions
   D. are *malleable*
   E. exhibit both negative and positive oxidation numbers

5. A water solution of _____ is classified as a base. 5.___
   A. CaO B. C_2H_5OH C. CO_2 D. CH_3COOH E. H_2O_2

6. An oxidizing agent 6.___
   A. always contains oxygen
   B. is always reduced in a redox reaction
   C. always donates electrons
   D. is a catalyst
   E. is an electron donor

7. An element forming an ion resembling the electronic structure of ${}_2He^4$ is 7.___
   A. ${}_3Li^6$ B. ${}_8O^{16}$ C. ${}_{11}Na^{23}$ D. ${}_{17}Cl^{35}$ E. ${}_{10}Ne^{20}$

8. A Lewis base 8.___
   A. must contain hydroxide ion
   B. is an electron pair donor
   C. is a substance that yields hydronium ion
   D. is one classification of $AlCl_3$
   E. is a proton acceptor

9. An example of a nonpolar solvent is 9.___
   A. H_2O B. C_2H_5OH C. CH_3Cl D. CCl_4 E. CO_2

10. Water is a good solvent for 10.___
A. C_2H_5OH B. CH_4 C. CS_2 D. CCl_4 E. Br_2

11. An element in the same family as sodium and *more* electro-positive than sodium is 11.___
A. rubidium B. oxygen C. chlorine
D. carbon E. fluorine

12. A strong acid that is *also* a strong oxidizing agent is 12.___
A. acetic B. phosphoric C. hydrochloric
D. nitric E. boric

13. Water has a *higher* boiling point than hydrogen sulfide because 13.___
A. water has the higher molecular weight
B. hydrogen sulfide ionizes
C. water molecules form aggregates by hydrogen bonding
D. water ionizes
E. water expands on freezing

14. The position of an element in the periodic table is determined by 14.___
A. electron configuration B. electronegativity
C. atomic volume D. atomic mass
E. number of neutrons

15. One mole of H_2SO_4 contains 15.___
A. 2 grams of hydrogen
B. Avogadros number of hydrogen atoms
C. 3 moles of atoms
D. 64 grams of sulfur
E. 7 atoms

16. One molecule of gas A reacts with three molecules of gas B to form 2 molecules of gas C and no other substance. Gas C contains _____ atoms. 16.___
A. 1 B. 2
C. 3 D. 4
E. all answers possible

17. Sodium perchlorate ($NaClO_4$) will decompose to give oxygen gas and sodium chloride. One mole sodium will yield _____ gram(s) oxygen. 17.___
A. 1 B. 16 C. 32 D. 48 E. 64

18. A flask contained .32 grams of oxygen gas. The same flask filled with an unknown gas at the same temperature and pressure contained .18 grams of gas. The molecular weight of the unknown gas is 18.___
A. 2 B. 9 C. 18 D. 22 E. 32

19. Two grams of a gas occupies 2.24 liters at standard temperature and pressure. The molecular weight of the gas is 19.___
A. 2.0 B. 16 C. 20 D. 22.4 E. 44.8

20. -218°C can be expressed as 20.___
A. -218°F B. 55°K C. 218°K D. 273°K E. 491°K

21. Of the following substances, the one with the HIGHEST boiling point is 21.___
A. He B. H_2 C. O_2 D. SO_2 E. S

22. A solution of a salt gives a violet flame test and a white precipitate when treated with barium chloride. The solution could contain 22.___
A. NaCl
B. Kl
C. K_2SO_4
D. $NH_4C_2H_3O_2$
E. KNO_3

23. An element that forms ionic solids with chlorine is 23.___
A. carbon
B. calcium
C. aluminum
D. phosphorus
E. hydrogen

24. When a liquid evaporates, one observes a cooling effect. This is accounted for by 24.___
A. the liquid being wet
B. energy being required to convert a liquid to a gas
C. movement of air
D. the conversion of liquid to gas being an exothermic process
E. energy being liberated when a liquid is converted to a gas

25. Given $H_2 + Br_2 = 2HBr + 17.2$ kilocalories, the heat liberated when .5 mole hydrogen reacts with .5 mole bromine is _____ kilocalorie(s). 25.___
A. .5 B. 1 C. 8.6 D. 17.2 E. 34.4

KEY (CORRECT ANSWERS)

1. D
2. D
3. D
4. E
5. A

6. B
7. A
8. B
9. D
10. A

11. A
12. D
13. C
14. A
15. A

16. D
17. E
18. C
19. C
20. B

21. E
22. C
23. B
24. B
25. C

TEST 3

DIRECTIONS: Each question or incomplete statement is followed by several suggested answers or completions. Select the one that BEST answers the question or completes the statement. *PRINT THE LETTER OF THE CORRECT ANSWER IN THE SPACE AT THE RIGHT.*

1. _____ is the STRONGEST acid. 1.____
   A. H_2S B. NH_4OH C. H_2SO_3 D. $HClO_4$ E. H_2PO_4

2. A substance which alters the speed of a chemical reaction but leaves the process unchanged is termed 2.____
   A. a cation
   B. a catalyst
   C. an activator
   D. a spectator ion
   E. nascent

3. The symbol CO represents 3.____
   A. carbon
   B. carbon monoxide
   C. cobalt
   D. cobalt oxide
   E. copper

4. A good example of a mixture is 4.____
   A. sodium perchlorate
   B. carbon monoxide
   C. hydrogen gas
   D. milk
   E. cane sugar

5. A compound containing only zinc and sulfur would be named zinc 5.____
   A. sulfide
   B. disulfide
   C. sulfite
   D. bisulfite
   E. sulfate

6. The formula of nitrous acid is 6.____
   A. H_2HO_4 B. HNO_3 C. HNO_2 D. HNO E. NH_4OH

7. Which of these substances is a nonelectrolyte when dissolved in water? 7.____
   A. Sugar
   B. Hydrogen chloride
   C. Sodium chloride
   D. Ammonia
   E. Acetic acid

8. One of the following electron arrangements represents that of an active nonmetal. 8.____
   A. 2 B. 2 8 C. 2 8 1 D. 2 8 3 E. 2 8 7

9. Oxygen may be prepared by 9.____
   A. heating a carbonate
   B. reacting an acid with a metal
   C. heating sodium chloride
   D. electrolysis of water
   E. reacting a metal oxide with water

10. A crystal of solid solute is added to a solution. The crystal does NOT change in mass. The solution was 10.___
   A. unsaturated   B. saturated
   C. supersaturated   D. dilute
   E. normal

11. The oxidation number of chlorine in $KClO_4$ is 11.___
   A. -1   B. +1   C. +4   D. +5   E. +7

12. The valence of aluminum in $Al_2(SO_4)_3$ is 12.___
   A. +1   B. +2   C. +3   D. +4   E. +6

13. The formula of potassium sulfate is 13.___
   A. PSO_3 B. Po_2SO_3 C. KSO_4 D. K_2SO_4 E. KS

14. Barium, which is in the same family as calcium, would be predicted to exhibit a valence of 14.___
   A. +1   B. +2   C. +3   D. +4   E. +5

15. One of the following elements would be MOST likely to be found as components of compounds in nature: 15.___
   A. Helium
   B. Platinum
   C. Potassium
   D. Gold
   E. All of these occur naturally as components of compounds

16. An electron arrangement that represents an inert gas is 16.___
   A. 2 8 2   B. 2 4   C. 2 8 7   D. 2 8 8   E. 2 8 3

17. The class of compounds to which H_3PO_4 belongs is 17.___
   A. acid   B. base   C. salt
   D. anhydride   E. phosphite

18. In gaseous hydrogen, the chemical bond is BEST classified as a(n) _____ bond. 18.___
   A. ionic   B. polar covalent
   C. covalent   D. hydrogen
   E. coordinate covalent

19. In solid potassium bromide, the chemical bond is BEST classified as a(n) _____ bond. 19.___
   A. ionic   B. polar covalent
   C. covalent   D. coordinate covalent
   E. bromine

20. A property that is characteristic of an electrovalent compound is 20.___
   A. low boiling point
   B. usually a gas at room temperature
   C. non-electrolyte
   D. high heat of vaporization
   E. all of these

21. A water would be classified as hard if it contained ______ ion. 21.___
A. calcium B. sodium C. chloride
D. potassium E. bicarbonate

22. $_1H^1$, $_1H^2$, $_1H^3$ are 22.___
A. isotopes B. allotropes
C. isomers D. chemically unrelated
E. hydrides

23. A base may be defined as 23.___
A. any strong electrolyte
B. a proton acceptor
C. an electron acceptor
D. any hydroxyl containing compound
E. ammonia

24. A piece of copper is coated with CuO. 24.___
A substance that would react with this compound without reacting with the copper is
A. NaOH B. $CaSO_4$ C. Na_2CO_3 D. HCl E. KNO_3

25. When acid is spilled, a nonhazardous material that should be used to neutralize the acid is 25.___
A. NaOH B. $NaHCO_3$ C. NaCl D. CH_3COOH E. HCl

KEY (CORRECT ANSWERS)

1. D
2. B
3. B
4. D
5. A

6. C
7. A
8. E
9. D
10. B

11. E
12. C
13. D
14. B
15. C

16. D
17. A
18. C
19. A
20. D

21. A
22. A
23. B
24. D
25. B

TEST 4

DIRECTIONS: Each question or incomplete statement is followed by several suggested answers or completions. Select the one that BEST answers the question or completes the statement. *PRINT THE LETTER OF THE CORRECT ANSWER IN THE SPACE AT THE RIGHT.*

1. 65° centigrade is equivalent to _____ ° absolute. 1.____
   A. 97 B. 117 C. 149 D. 273 E. 338

2. A two molar solution of phosphoric acid, molecular weight 98, contains _____ grams solute per liter. 2.____
   A. 2 B. 49 C. 98 D. 196 E. 392

3. Carbon dioxide may be prepared by reacting sodium carbonate with 3.____
   A. sodium oxide
   B. hydrochloric acid
   C. water
   D. potassium hydroxide
   E. calcium chloride

4. A sugar solution is boiling.
   As the water boils away, the temperature 4.____
   A. is 100 degrees
   B. remains constant
   C. decreases
   D. increases
   E. depends on source of heat

5. A chemical reaction is started by heating the reacting substances. The reaction continues when the heat is removed. 5.____
   This indicates the reaction
   A. is exothermic
   B. is endothermic
   C. yields carbon dioxide
   D. is very slow
   E. is at equilibrium

6. The predicted compound of element number seven which we shall represent as Y with hydrogen would be 6.____
   A. HY
   B. H_2Y
   C. H_3Y
   D. H_4Y
   E. none of these

7. The predicted compound of element number eleven with chlorine would be 7.____
   A. an electric conductor in the molten state
   B. predicted to exist as individual molecules
   C. a gas at room temperature
   D. an acid
   E. a nonelectrolyte in aqueous solutions

8. The type of chemical bond that is considered to account for the unusual properties of water is a(n) _____ bond. 8.___
   A. ionic
   B. covalent
   C. hydrogen
   D. coordinate covalent
   E. ion dipole

9. Zinc metal is placed in concentrated sulfuric acid. It would 9.___
   A. react violently
   B. react at a very slow rate
   C. NOT react as sulfuric acid as an oxidizing acid
   D. react only at low temperatures
   E. react more rapidly than in 6 normal acid

Questions 10-14.

DIRECTIONS: For the reaction $2H_3PO_4 + 3Zn \rightarrow Zn_3(PO_4)_2 + 3H_2$ using the atomic weights H = 1, O = 16, P = 31, Zn = 65.4.

10. The molecular weight of H_3PO_4, phosphoric acid, is 10.___
    A. 2 B. 48 C. 96 D. 98 E. 196

11. Four gram atoms of zinc will yield 11.___
    A. four moles hydrogen
    B. four grams hydrogen
    C. four liters hydrogen at STP
    D. eight moles hydrogen
    E. eight liters hydrogen at STP

12. 196 grams phosphoric acid will yield 12.___
    A. 1 gram hydrogen
    B. 3 grams hydrogen
    C. 6 grams hydrogen
    D. 3 liters hydrogen at STP
    E. 6 liters hydrogen at STP

13. Four moles of phosphoric acid will yield 13.___
    A. 1 mole zinc phosphate
    B. 2 moles zinc phosphate
    C. 4 moles zinc phosphate
    D. 3 moles hydrogen
    E. 22.4 liters of hydrogen at STP

14. 392 grams phosphoric acid will yield _____ liter(s) hydrogen at STP. 14.___
    A. 1 B. 3 C. 22.4 D. 67.2 E. 134.4

Questions 15-18.

DIRECTIONS: For the reaction $H_2 + Cl_2 \rightarrow 2HCl$ using atomic weights $H = 1$ and $Cl = 35.5$.

15. Four liters of hydrogen would react with _____ liters chlorine. 15.___
A. 2 B. 4 C. 35.5 D. 89.6 E. 142

16. 22.4 liters hydrogen at STP would yield _____ grams HCl. 16.___
A. 2
B. 36.5
C. 73
D. 877
E. correct answer not given

17. 35.5 grams of chlorine would yield _____ grams HCl. 17.___
A. 2 B. 35.5 C. 36.5 D. 71 E. 73

18. Two moles of HCl weigh _____ grams. 18.___
A. 2 B. 4 C. 36.5 D. 73 E. 219

19. The weight of 2.24 liters of oxygen gas at STP is _____ grams. 19.___
A. 2.24 B. 1.6 C. 3.2 D. 16.0 E. 32.0

20. The molecular weight of a gas that exhibits a density of 3.2 grams per liter at STP is 20.___
A. 3.2 B. 2.24 C. 32 D. 71.5 E. 7.0

21. The *approximate* volume of 3 moles of methane gas (molecular weight 20) at STP is _____ liters. 21.___
A. 3 B. 20 C. 22.4 D. 67 E. 213

22. The volume of a gas halves at constant pressure. The temperature 22.___
A. remains constant
B. doubles in degrees absolute
C. is decreased to 1/2 the original value in degrees absolute
D. was 273 degrees absolute
E. increases to 4 times the original value in degrees centigrade

23. The numerical value of the vapor pressure of water at 100° centigrade is _____ mm. of Hg. 23.___
A. 22.4 B. 100 C. 224 D. 760 E. 1000

24. 100 grams of ice at 0°C is melted and converted to liquid water at 20°C. 24.___
It requires _____ calories.
A. 20 B. 80 C. 2000 D. 8000 E. 10,000

25. A solid changes to a gas and then reforms solid. This represents 25.____
A. precipitation
B. condensation
C. evaporation
D. sublimation
E. homogenization

KEY (CORRECT ANSWERS)

1. E
2. D
3. B
4. D
5. A
6. C
7. A
8. C
9. B
10. D
11. A
12. C
13. B
14. E
15. B
16. C
17. C
18. D
19. C
20. D
21. D
22. B
23. D
24. E
25. D

EXAMINATION SECTION

TEST 1

DIRECTIONS: Each question or incomplete statement is followed by several suggested answers or completions. Select the one that BEST answers the question or completes the statement. *PRINT THE LETTER OF THE CORRECT ANSWER IN THE SPACE AT THE RIGHT.*

1. Analysis of 1 mg. samples is termed 1.___
   A. microanalysis B. macroanalysis
   C. meganalysis D. semimicroanalysis

2. Lambert's law is *generally* associated with _____ methods of analysis. 2.___
   A. electrical B. mechanical
   C. optical D. thermal

3. Of the following substances, the one classified as an aliphatic compound is _____ acid. 3.___
   A. benzoic B. oxalic C. phthalic D. salicylic

4. Of the following substances, the one classified as an aromatic compound is _____ acid. 4.___
   A. acetic B. carbolic C. carbonic D. citric

5. The one of the following carbohydrates classified as an polysaccharide is 5.___
   A. cellulose B. fructose C. glucose D. maltose

6. Of the following, the one classified as an amino acid is _____ acid. 6.___
   A. glutamic B. lactic C. oleic D. oxalic

7. Of the following compounds, the one which is a member of the paraffin series of hydrocarbons is 7.___
   A. acetylene B. cyclohexane
   C. ethane D. ethylene

8. The carbohydrate obtained from sugar cane is called 8.___
   A. dextrose B. lactose C. starch D. sucrose

9. The name of the enzyme that catalyzes the decomposition of starch into sugar is 9.___
   A. amylase B. lipase C. pepsin D. rennin

10. The one of the following which is a form of electromagnetic radiation is _____ rays. 10.___
    A. alpha B. beta C. cathode D. gamma

11. The unit of electric current is the 11.___
    A. ampere B. coulomb C. joule D. volt

12. In centrifuging, the rate of sedimentation varies inversely with the 12.___
   A. centrifugal force
   B. difference in density between the solid and that of the solution
   C. size of the particle
   D. viscosity of the solution

13. The symbol Å is used to designate a unit of 13.___
   A. charge
   B. length
   C. mass
   D. specific gravity

14. In the electrolysis of water, the gas liberated at the anode is 14.___
   A. CO B. CO_2 C. H_2 D. O_2

15. The use of a safety bottle is *generally* recommended in 15.___
   A. colorimetric determinations
   B. decantations of small volumes
   C. electrolytic precipitations
   D. filtrations by suction

16. In colimetry, the name of tubes frequently used for comparison of colors is _____ tubes. 16.___
   A. Inner B. Litmus C. Nessler D. Wolff

17. SO_2 is the anhydride of 17.___
   A. H_2SO_3 B. H_2SO_4 C. $H_2S_2O_3$ D. $H_2S_2O_7$

18. Of the following types of glass, the one with the SMALLEST coefficient of thermal expansion is _____ glass. 18.___
   A. pyrex
   B. safety
   C. soda-lime
   D. soft

19. The product, volts x amperes, is equal to 19.___
   A. faradays B. joules C. ohms D. watts

20. The pH of a solution whose (H^+) is 1×10^{-4} mole per liter is 20.___
   A. 4 B. 5 C. 6 D. 7

Questions 21-30.

DIRECTIONS: Column I lists terms, each of which can be properly matched with one of the pieces of laboratory equipment in Column II. For each item of Column I, write in the space at the right the letter in front of the option in Column II with which it is generally associated.

COLUMN I	COLUMN II	
21. Daniel	A. balance	21.___
22. Duboscq	B. beaker	22.___
23. Erlenmeyer	C. burner	23.___
24. Gooch	D. cell	24.___
25. Griffin	E. colorimeter	25.___
26. Jones	F. condenser	26.___
27. Liebig	G. crucible	27.___
28. Meker	H. flask	28.___
29. Plattner	I. mortar	29.___
30. Westphal	J. reductor	30.___
	K. tube	
	L. viscosimeter	

KEY (CORRECT ANSWERS)

1. A	11. A	21. D
2. C	12. D	22. E
3. B	13. B	23. H
4. B	14. D	24. G
5. A	15. D	25. B
6. A	16. C	26. J
7. C	17. A	27. F
8. D	18. A	28. C
9. A	19. D	29. I
10. D	20. A	30. A

TEST 2

DIRECTIONS: Each question or incomplete statement is followed by several suggested answers or completions. Select the one that BEST answers the question or completes the statement. *PRINT THE LETTER OF THE CORRECT ANSWER IN THE SPACE AT THE RIGHT.*

1. Of the following, the substance whose water solution is acid is 1.___
   A. CH_3OH B. $NaHCO_3$ C. NH_3 D. $(NH_4)HSO_4$

2. The instrument *generally* used to measure electric current is the 2.___
   A. ammeter B. coulometer
   C. potentiometer D. voltmeter

3. Of the following substances, the one which is LEAST soluble in water is 3.___
   A. $CuSO_4$ B. K_2SO_4 C. $MgSO_4$ D. $SrSO_4$

4. In qualitative analysis, potassium thiocyanate is used to test for 4.___
   A. Al^{+++} B. Cu^{++} C. Fe^{+++} D. Ni^{++}

5. Dimethylglyoxime is used as a specific reagent for the detection of 5.___
   A. Ca^{++} B. Ni^{++} C. NO_3^- D. SO_4^{--}

6. One liter of a 2 M HCl solution will *readily* dissolve 10 grams of 6.___
   A. Ag_2S B. CuS C. S D. ZnS

7. A colorless solution is treated with dilute HCl giving a white precipitate. When the precipitate is treated with 6 M NH_3, the white precipitate turns black. The original solution must have contained 7.___
   A. Ag^+ B. Hg^{++} C. Hg_2^{++} D. Pb^{++}

8. A mixture of $BaCl_2$, LiCl, NaCl, and NH_4Cl is heated in an evaporating dish over a strong Bunsen flame. The cloud of white vapor which is observed is due to the 8.___
   A. $BaCl_2$ B. LiCl C. NaCl D. NH_4Cl

9. A violet color obtained in a flame test indicates the presence of 9.___
   A. copper B. lead C. potassium D. sodium

10. A solution of $KMnO_4$ may be standardized by titration against 10.___
    A. oxalic acid B. potassium acid phthalate
    C. sodium bicarbonate D. sulfuric acid

11. A solution of sodium hydroxide may be standardized by titration against 11.___
    A. barium diphenylamine sulfonate
    B. potassium acid phthalate
    C. sodium carbonate
    D. sodium thiosulfate

12. Cellulose is a 12.___
    A. mineral deposit in plant cells
    B. mixture of proteins
    C. polysaccharide
    D. synthetic rubber

13. The heat of combustion of an organic compound is *generally* measured by means of a 13.___
    A. bomb calorimeter
    B. Carius tube
    C. pyrometer
    D. Victor Meyer apparatus

14. The amount of nitrogen in an organic compound may be measured by means of 14.___
    A. an emission spectroscope
    B. a polarimeter
    C. the Hell-Volhard-Zelinsky reaction
    D. the Kjeldahl method

15. Small amounts of soluble solid impurities in solutions of organic compounds are *frequently* removed by boiling with 15.___
    A. activated charcoal
    B. boiling stones
    C. glass beads
    D. small scraps of rubber tubing

16. Enzymes are 16.___
    A. an essential part of the diet
    B. biological catalysts
    C. hormones
    D. vitamins

17. In quantitative analysis, a precipitate of silver chloride is *generally* separated from the supernatant solution by 17.___
    A. a Gooch crucible
    B. a powder funnel
    C. a separatory funnel
    D. mixing with strong NH_3 solution and then decanting

18. The amount of rise of a liquid in a capillary is used to measure 18.___
    A. conductivity
    B. diffusion coefficient
    C. surface tension
    D. viscosity

19. The molecular weight of an *unknown* may be determined by dissolving a definite amount of *unknown* in a measured amount of solvent and determining the 19.___
   A. conductivity
   B. freezing point depression
   C. temperature coefficient of conductivity
   D. temperature coefficient of surface tension

20. A current of 10 amperes at 110 volts corresponds to ______ watts. 20.___
   A. 0.091 B. 11 C. 1100 D. 11,000

21. A diffusion pump is *generally* used 21.___
   A. for suction filtration
   B. to maintain the level of a water bath
   C. to obtain a high pressure
   D. to obtain a high vacuum

22. An endothermic reaction is one which 22.___
   A. absorbs heat
   B. liberates heat
   C. takes place only at low temperature
   D. takes place only at high temperature

23. A voltage of 110 volts flowing through a 110 ohm resistor produces 23.___
   A. 1 watt
   B. 110 watts
   C. 12,100 watts
   D. a wattage depending upon the time the current is allowed to flow

24. Stannous chloride is 24.___
   A. a reducing agent
   B. an oxidizing agent
   C. used for galvanizing iron
   D. very insoluble in water

25. The atomic weight of an element is determined by the 25.___
   A. number of neutrons in the nucleus
   B. number of planetary electrons
   C. number of protons in the nucleus
   D. sum of the number of protons and neutrons in the nucleus

KEY (CORRECT ANSWERS)

1. D	6. D	11. B	16. B	21. D
2. A	7. C	12. C	17. A	22. A
3. D	8. D	13. A	18. C	23. B
4. C	9. C	14. D	19. B	24. A
5. B	10. A	15. A	20. C	25. D

EXAMINATION SECTION
TEST 1

DIRECTIONS: Each question or incomplete statement is followed by several suggested answers or completions. Select the one that BEST answers the question or completes the statement. *PRINT THE LETTER OF THE CORRECT ANSWER IN THE SPACE AT THE RIGHT.*

The following list of atomic weights may be referred to in solving problems involving computations:

Calcium	40.1	Hydrogen	1.0	Potassium	39.1
Carbon	12.0	Iodine	126.9	Silver	107.9
Chlorine	35.5	Nitrogen	14.0	Sodium	23.0
Chromium	52.0	Oxygen	16.0	Sulfur	32.0

1. A catalyst is a substance which 1.___
   A. increases the rate of the reaction
   B. is made of platinum
   C. pushes the point of equilibrium farther towards the products
   D. slows up side reactions

2. One liter is APPROXIMATELY 2.___
   A. 1 pint B. 2 pints C. 2 quarts D. 1 gallon

3. An electrolyte is a compound whose solution 3.___
   A. decomposes water
   B. has an electrical potential
   C. has the property of electrical conduction
   D. moves in an electrical field

4. Oxides of non-metals when dissolved in water 4.___
   A. evolve hydrogen B. evolve oxygen
   C. form acids D. form bases

5. Of the following metals, the one which is MOST active chemically is 5.___
   A. copper B. iron C. nickel D. zinc

6. When the temperature is increased, the rate of a simple chemical reaction is 6.___
   A. decreased
   B. increased
   C. sometimes increased and sometimes decreased
   D. unaffected

7. The pressure in a freshly filled standard cylinder of oxygen is APPROXIMATELY 7.___
   A. 100 lbs/in^2 B. 400 lbs/in^2
   C. 2000 lbs/in^2 D. 400 atmospheres

8. Of the following substances, the one which is LEAST soluble in water at 25°C is 8.___
A. ferric sulfate B. lead nitrate
C. mercuric chloride D. silver iodide

9. Of the following solids, the one with the HIGHEST melting point is 9.___
A. lead B. silver
C. silver nitrate D. tungsten

10. An accurate estimate of the number of protons in an atomic nucleus is obtained by taking the atomic 10.___
A. number
B. number plus the atomic weight
C. weight
D. weight minus the atomic number

11. Of the following, the MOST electropositive element is 11.___
A. arsenic B. copper C. mercury D. rubidium

12. If 100 grams of a substance requires 100 calories to raise its temperature 10°C, the specific heat of the substance is 12.___
A. .01 B. 0.1 C. 1.0 D. 10.0

13. The volume occupied by 200 grams of a solution whose density is 0.92 is MOST NEARLY ____ ml. 13.___
A. 92 B. 109.5 C. 134 D. 217.5

14. Of the following, the one which would be the MOST appropriate for keeping a dry ice and acetone mixture for several hours is the ____ flask. 14.___
A. Claisen B. Dewar
C. Ehrlenmeyer D. Florence

15. A solution of KNO_3 containing 303 grams in one liter of solution is APPROXIMATELY ____ molar. 15.___
A. 1 B. 3 C. 10 D. 12

16. A violet color obtained in a flame test indicates the presence of 16.___
A. barium B. mercury C. potassium D. sodium

17. One gram of an *unknown* metal reacts completely with dilute hydrochloric acid to release 112 ml of hydrogen (measured at STP). 17.___
The equivalent weight of the metal is ____ gm.
A. 50 B. 100 C. 200 D. 400

18. An unknown white solid dissolves in 0.3M HCl to give a clear solution. When H_2S is bubbled through this solution, no reaction is observed. 18.___
The solution, therefore, CANNOT contain
A. NH_4^+ B. BA^{++} C. Hg^{++} D. $SO_4^{=}$

19. If 50 ml of a solution of H_2SO_4 exactly neutralizes 25 ml of a 1.00 N solution of NaOH, the normality of the H_2SO_4 solution is 19.___
A. 0.50 B. .25 C. 1.0 D. 2.0

20. A Kjeldahl analysis can be used to determine the percentage of 20.___
A. acetone in urine B. sugar in urine
C. nitrogen in protein D. sulfate in water

21. Potassium thiocyanate solution gives an intense red solution with 21.___
A. Fe^{++} B. Ag^{+}
C. Mn^{++} D. none of the above

22. The concentration of silver chromate in a saturated solution is approximately 1.3×10^{-4}M. 22.___
The solubility product of silver chromate is
A. 9×10^{-12} B. 2.2×10^{-12}
C. 1.7×10^{-8} D. 3.9×10^{-4}

23. Of the following, the STRONGEST oxidizing agent is sodium 23.___
A. dichromate and sulfuric acid
B. nitrate and sodium hydroxide
C. thiocyanate and ammonia
D. thiosulfate and hydrochloric acid

24. The name of the instrument used for the precise determination of electrical voltage is the 24.___
A. ammeter B. ohmmeter
C. potentiometer D. Wheatstone bridge

25. Which of the following will dissolve the MOST barium hydroxide? 25.___
A. 2 ml of 1.0M nitric acid
B. 10 ml of 0.2M perchloric acid
C. 1.5 ml of 0.1M sulfuric acid
D. 25 ml of 0.1M hydrochloric acid

26. An unknown white solid dissolves in 0.3M HCl to give a clear solution. 26.___
It, therefore, CANNOT contain ____ ions.
A. calcium B. mercurous C. arsenic D. rubidium

27. Of the following solids, the one which is MOST readily soluble in 6M NaOH is 27.___
A. barium sulfate B. ferric chloride
C. hydrated zinc hydroxide D. mercuric chloride

28. An *unknown* metal dissolves readily in dilute nitric acid to give a colorless solution (ignore color due to nitric acid). This solution was evaporated to dryness, and the residue was then taken up in water. Treatment of this aqueous solution with thioacetamide (or hydrogen sulfide) gave a yellow precipitate. 28.___
The *unknown* metal was PROBABLY
A. Cd B. Ni C. Pt D. Zn

29. The one of the following titrations in which phenolphthalein is useful as an indicator is 29.___
A. acetic acid with ammonia
B. acetic acid with sodium hydroxide
C. ammonia with hydrochloric acid
D. sodium hydroxide with hydrochloric acid

30. The series in which each element will displace from aqueous solution those following it is 30.___
A. Ag, Hg, H, N, Cu
B. K, Fe, H, Hg, Pt
C. K, Zn, Pb, Pt, H
D. Na, H, Zn, Pb, Pt

31. A pH of 2 means that the H^+ is equal to 31.___
A. 2
B. 10^2
C. 10^{-2}
D. e^{-2}

32. Raoult's Law 32.___
A. gives the vapor pressure of a liquid as a function of temperature
B. relates the specific heat and atomic weight of certain solids
C. relates the vapor pressure and concentration of one component of a binary solution
D. tells which way an equilibrium changes with the temperature

33. A *perfect* gas is a gas which 33.___
A. does not change pressure with a change in temperature if the volume is kept constant
B. follows van der Waals' equation exactly
C. forms no chemical compounds
D. has one equation of state, PV = NRT

34. The vapor pressure of liquids 34.___
A. decreases with increasing temperature
B. does not change with increasing temperature
C. increases with increasing temperature
D. sometimes increases and sometimes decreases with temperature

35. The nucleus of a chloride ion contains _____ the chlorine atom from which it is derived. 35.___
A. one more electron than
B. one more neutron than
C. one more proton than
D. the same number of protons and neutrons as

36. Analysis of a compound gave 14.1%C, 2.4%H, and 83.5%Cl. The compound was PROBABLY 36.___
A. carbon tetrachloride
B. chloroform
C. chloromethane
D. dichloromethane

37. Optical isomers are also called 37.___
A. allotropic forms
B. enantiomorphs
C. geometrical isomers
D. peritectics

38. A soap solution in water foams readily because soap ____ of water. 38.____
   A. *lowers* the surface tension
   B. *raises* the viscosity
   C. *lowers* the viscosity
   D. *raises* the surface tension

39. The extraction of I_2 from 100 ml of a water solution is performed MOST efficiently by 39.____
   A. one 50 ml portion of CCl_4
   B. one 50 ml portion of ethanol
   C. five 10 ml portions of CCl_4
   D. five 10 ml portions of ethanol

40. A reliable way to construct a straight line through a set of experimentally obtained points on a graph is the 40.____
   A. latin square
   B. method of convergent slopes
   C. method of least squares
   D. method of random samples

KEY (CORRECT ANSWERS)

1. A	11. D	21. D	31. C
2. B	12. B	22. A	32. C
3. C	13. D	23. A	33. D
4. C	14. B	24. C	34. C
5. D	15. B	25. D	35. D
6. B	16. C	26. B	36. D
7. C	17. B	27. C	37. B
8. D	18. C	28. A	38. A
9. D	19. A	29. B	39. C
10. A	20. C	30. B	40. C

TEST 2

DIRECTIONS: Each question or incomplete statement is followed by several suggested answers or completions. Select the one that BEST answers the question or completes the statement. *PRINT THE LETTER OF THE CORRECT ANSWER IN THE SPACE AT THE RIGHT.*

1. A temperature of 450°C is MOST readily measured with a 1.___
   A. iron-constantan thermocouple
   B. mercury thermometer
   C. adiabatic calorimeter
   D. spinthariscope

2. Of the following, the BEST drying agent is 2.___
   A. anhydrous potassium bromide
   B. concentrated HCl
   C. hydrated sodium sulfate
   D. P_2O_5

3. In order to make 50 ml of approximately 0.3M HCl, enough water to make 100 ml of solution should be added to 3.___
   A. 0.36 ml of 12M HCl
   B. 2.5 ml of 6M HCl
   C. 5.0 ml of 12M HCl
   D. 3.6 ml of 12M HCl

4. The purity of distilled water used in the laboratory may be estimated MOST readily by 4.___
   A. analyzing for likely impurities
   B. measuring the absorption spectrum
   C. measuring the boiling point
   D. measuring the electrical conductivity

5. The density of a liquid can be readily measured with a(n) 5.___
   A. densitometer
   B. galvanometer
   C. pycnometer
   D. isoteniscope

6. The average reading error on a standard 50 ml burette is APPROXIMATELY ____ ml. 6.___
   A. .002
   B. .02
   C. .2
   D. 2

7. The average weighing error on the standard analytical balance is MOST NEARLY 7.___
   A. 0.2 milligrams
   B. 20 milligrams
   C. .02 grams
   D. 0.2 grams

8. In the preparation of potassium permanganate solution, a period of storage is recommended before standardization to allow time for 8.___
   A. diffusion of the permanganate through the solution
   B. escape of air from the solution
   C. oxidation of accidentally included organic impurities such as dust
   D. reduction of accidentally included organic impurities such as dust

9. The recommended method for drying ethanol is 9.___
   A. by fractional distillation
   B. in an oven
   C. over anhydrous Na_2SO_4
   D. over sodium

10. Indicators for pH are GENERALLY 10.___
   A. acid catalysts
   B. easily oxidizable compounds
   C. easily reducible compounds
   D. weak acids or bases

11. Of the following, the one with the LOWEST freezing point is 11.___
   A. acetic acid B. benzene
   C. mercury D. water

12. Of the following, the one with the HIGHEST melting point is 12.___
   A. Au B. Al_2O_3 C. KCl D. Na

13. Of the following gases, the one with the HIGHEST boiling point is 13.___
   A. NH_3 B. SO_2 C. N_2 D. He

14. As a lead storage battery is discharged, 14.___
   A. sulfuric acid should be added to the cell
   B. the density of the liquid in the cell decreases
   C. the density of the liquid in the cell increases
   D. the density of the liquid in the cell is unchanged

15. Of the following substances, the one which dissolves to the LEAST extent in 1M HCl at 25°C is 15.___
   A. copper sulfide B. silver carbonate
   C. sodium sulfate D. zinc hydroxide

16. The volumetric determination of copper USUALLY employs titration with 16.___
   A. ammonia B. potassium iodide
   C. sodium hydroxide D. sodium sulfide

17. A conductometric titration would be PARTICULARLY useful for determining 17.___
   A. the electrical properties of a transformer coil
   B. non-protein nitrogen in blood
   C. oxidizing strength of hydrogen peroxide
   D. total acidity of a highly colored solution

18. The unit in which viscosities are often reported is the 18.___
   A. millipoise B. coulomb C. dyne D. amagat

19. The component part of an instrument widely used for measuring pH is 19.___
   A. dropping mercury B. glass electrode
   C. protective electrode D. sacrificial electrode

20. The freezing point of a dilute solution of any salt in water is 20. ____
A. always 0°C
B. always higher than 0°C
C. always lower than 0°C
D. sometimes higher and sometimes lower than 0°C

21. A protein consists of 21. ____
A. C and H *only* B. C, H, and N *only*
C. C, H, and O *only* D. at least C, H, N, and O

22. An aldehyde may be produced by a mild oxidation of a 22. ____
A. phenol B. primary alcohol
C. secondary alcohol D. tertiary alcohol

23. Amino acids are formed in the hydrolysis of 23. ____
A. carbohydrates B. fats
C. hydrocarbons D. proteins

24. The one of the following elements which is NOT presently known to be usable as a fuel in a nuclear fission reactor is 24. ____
A. hydrogen B. plutonium C. thorium D. uranium

25. Of the following compounds, the one which possesses optical isomers is 25. ____

A. $CH_2 = CH - CH_2 - CH_3$ B. $CH_3 - CHCl - CH_2 - CH_3$

C. $H - \overset{\overset{O}{\|}}{C} - NH_2$ D. 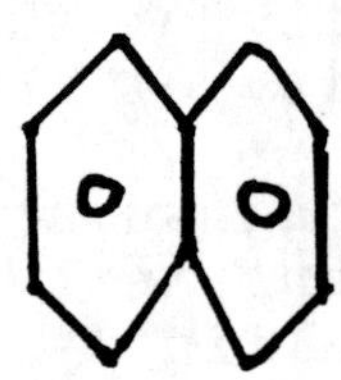 $- O - CH_3$

26. A solution of one metal in another is termed a(n) 26. ____
A. alloy B. cermet
C. double displacement D. twinned crystal

27. The molecular weight of a substance can be calculated from a measurement of its(the) 27. ____
A. density in the liquid state
B. freezing point
C. vapor pressure
D. density of its vapor

28. Elements with the same number of protons in the nucleus but different numbers of neutrons are called 28. ____
A. isentropes B. isomers
C. isomorphs D. isotopes

29. The quality of a precipitate for analytical purposes is often improved by 29. ____
A. co-precipitation B. digestion
C. double filtration D. salting out

30. In order to make a bend in 20 mm pyrex tubing, one should use a(n) 30.___
  A. blue Bunsen flame B. oxygen-gas torch
  C. oxygen-hydrogen torch D. yellow Bunsen flame

31. If 1M NaOH is slowly poured into a beaker containing 0.1M $AgNO_3$, 31.___
  A. a brown precipitate is formed which then redissolves
  B. a light brown precipitate is formed which then turns black
  C. a light brown precipitate is formed which then turns red
  D. the solution gets very warm and evolves brown fumes which have a pungent odor

32. If a piece of Zn is dropped into a beaker containing 6M NaOH, 32.___
  A. a black film forms over the surface
  B. a white crust forms over the metal surface
  C. gas is evolved
  D. no significant change is seen

33. A crude instrument for measuring surface tension could be readily improvised from a(n) 33.___
  A. graduated cylinder and a manometer
  B. piece of capillary tubing and a ruler
  C. thermos bottle and a thermometer
  D. analytical balance and a pair of pliers

34. The condition of a small opaque fuse is readily checked with a(n) 34.___
  A. ammeter B. ohmmeter
  C. potentiometer D. voltmeter

35. A device often used to vary an electrical current is the 35.___
  A. coulometer B. polarograph
  C. rheostat D. vacuum tube voltmeter

36. A neutral solution will result upon mixing equal volumes of 0.1 N solutions of 36.___
  A. hydrogen sulfide and sodium hydroxide
  B. nitric acid and calcium hydroxide
  C. phosphoric acid and sodium hydroxide
  D. sodium chloride and ammonium chloride

37. Potassium permanganate may be used in analytical chemistry for the direct determination of 37.___
  A. arsenious compounds, carbonates, and ferric chloride
  B. chloride, oxalate, and metals
  C. ferrocyanide, oxalate, and sulfite
  D. ferricyanide, sulfite, and thiosulfate

38. The refractive indices of liquids are GENERALLY between ____ and ____. 38.___
  A. 0; 0.1 B. 0; 1.0 C. 1; 10 D. 10; 100

39. A photographic developer is a 39.___
 A. mild oxidizing agent B. mild reducing agent
 C. solvent for silver D. solvent for silver halides

40. Comparing solution #1, containing exactly one mole of potassium bromide and one mole of calcium nitrate, with solution #2, containing one mole of calcium bromide and one mole of potassium nitrate, the solutions 40.___
 A. differ in bromide content
 B. differ in calcium content
 C. differ in potassium content
 D. are identical

KEY (CORRECT ANSWERS)

1. A	11. C	21. D	31. B
2. D	12. B	22. B	32. C
3. B	13. B	23. D	33. B
4. D	14. B	24. A	34. B
5. C	15. A	25. B	35. C
6. B	16. B	26. A	36. B
7. A	17. D	27. D	37. C
8. C	18. A	28. D	38. C
9. C	19. B	29. B	39. B
10. D	20. C	30. B	40. A

TEST 3

DIRECTIONS: Each question or incomplete statement is followed by several suggested answers or completions. Select the one that BEST answers the question or completes the statement. *PRINT THE LETTER OF THE CORRECT ANSWER IN THE SPACE AT THE RIGHT.*

1. The term synonymous with biological catalyst is 1.___
   A. enzyme B. hormone C. vaccine
   D. virus E. vitamin

2. The name MOST closely associated with the Periodic Table is 2.___
   A. Arrhenius B. Born C. Einstein
   D. Mendelejeff E. Pauling

3. A solution of one metal in another is termed a(n) 3.___
   A. alloy B. cermet
   C. double displacement D. solvolysis
   E. twinned crystal

4. The branch of chemistry that is PRIMARILY concerned with the rate of chemical reactions is called 4.___
   A. cryogenics B. kinaesthetics C. kinetics
   D. mechanics E. thermodynamics

5. Isotopes are atoms of elements which differ in 5.___
   A. atomic weight and extranuclear charge
   B. atomic weight and nuclear charge
   C. atomic weight but not in nuclear charge
   D. nuclear charge but not in extranuclear charge
   E. valency

6. The one of the following elements which is NOT presently known to be usable as a fuel in a nuclear fission reactor is 6.___
   A. actinium B. hydrogen C. plutonium
   D. thorium E. uranium

7. An electrolyte is a compound whose solution 7.___
   A. decomposes water
   B. does electrical work
   C. has an electrical potential
   D. has the property of electrical conduction
   E. moves in the electrical field

8. The *inert* gases do NOT form compounds because 8.___
   A. all of the electron shells are completed
   B. their kinetic energies are so great that they cannot approach any other atom
   C. their sizes are too great to form compounds
   D. they have even numbers of electrons
   E. they have equal numbers of protons and electrons in their structure

9. The number of positive charges in the nucleus of a neutral element and the number of electrons in the orbits about it differ by 9.____
A. 0 B. 1 C. 2 D. 3 E. 4

10. A common unit of radioactivity is the 10.____
A. angstrom B. curie C. geiger
D. radian E. standard rat unit

11. Water gas is a mixture of 11.____
A. CO, H_2 and CH_4 B. CO and H_2O
C. CO, H_2O and H_2 D. CO_2 and H_2
E. CO_2 and H_2O

12. Ordinary soft solder is composed PRINCIPALLY of 12.____
A. antimony and bismuth B. antimony and tin
C. bismuth and lead D. bismuth and tin
E. lead and tin

13. The MOST active halogen is 13.____
A. astatine B. bromine C. chlorine
D. fluorine E. iodine

14. The elements F, Cl, Br, and I form compounds with all elements in groups I through VII of the periodic table. The compounds form less covalent and more electrovalent bonds as the group number of the other elements comes closer to 14.____
A. I B. III C. IV D. V E. VII

15. Of the following, the one with the LOWEST freezing point is 15.____
A. acetic acid B. acetone C. benzene
D. mercury E. water

16. Of the following groups of elements, the one whose oxides all form alkalies when dissolved in water is 16.____
A. Al, C, Fe, and S B. B, C, P, and S
C. B, Ca, Mg, and P D. Ba, Cl, Na, and S
E. Ca, Mg, K, and Na

17. Of the following acids, the one whose dissociation constant is nearest to 10^{-5} is _____ acid. 17.____
A. acetic B. carbonic C. hydrochloric
D. nitric E. sulfuric

18. Of the following salts, the one with the GREATEST solubility in water is _____ chloride. 18.____
A. cupric B. cuprous C. lead
D. mercurous E. silver

19. Of the following, the STRONGEST oxidizing agent is sodium 19.____
A. dichromate and sulfuric acid
B. iodide and sodium hydroxide
C. nitrate and sodium hydroxide
D. thiocyanate and ammonia
E. thiosulfate and hydrochloric acid

20. The one of the following which is a sulfite is 20.____
A. $NaHSO_4$ B. Na_2S C. Na_2SO_3 D. Na_2SO_4 E. $Na_2S_2O_7$

21. Of the following groups of elements, the one whose oxides all form acids when dissolved in water is 21.____
A. Al, B, Fe, and S
B. B, C, P, and S
C. B, Be, P, and S
D. C, Ca, N, and S
E. Ca, K, Mg, and Na

22. The one of the following statements regarding the use of the analytical balance that is INCORRECT is 22.____
A. analytical samples should be placed directly on the balance pan
B. the beam and stirrups should be taken off their knife edges between weighings
C. the empty balance need not be adjusted to *zero* before use
D. the object to be weighed should be at the same temperature as the balance
E. weights should be handled only with forceps

23. The one of the following sets of glassware that would be MOST useful for preparing a standard solution of potassium acid phthalate is weighing bottle, 23.____
A. graduated cylinder, funnel, glass rod, and beaker
B. graduated cylinder, funnel, glass rod, and transfer pipette
C. volumetric flask, funnel, glass rod, and beaker
D. volumetric flask, funnel, glass rod, and transfer pipette
E. volumetric flask, glass rod, beaker, and transfer pipette

24. Reagent grade concentrated hydrochloric acid is APPROXIMATELY 24.____
A. 6M B. 12M C. 18M D. 24M E. 36M

25. A common method for preparing a standardized 0.1N solution of NaOH is to 25.____
A. dilute a clarified 50% (saturated) solution of sodium hydroxide with boiled water and standardize against crystalline potassium acid phthalate
B. prepare a 0.1N solution of sodium bicarbonate and boil out the CO_2
C. prepare a 0.1N solution of sodium carbonate and precipitate the carbonate with excess barium hydroxide
D. weigh out exactly 2.30 g of sodium and dissolve it in one liter of water
E. weigh out exactly 4.00 g of sodium hydroxide and dissolve it in a liter of water

26. When one drop of 0.1M potassium thiocyanate solution is added to 10 ml of 0.1M ferric chloride, the color of the mixture is 26.____
A. blue
B. green
C. nearly colorless
D. orange-yellow
E. red

27. When a gas volume is measured over water, a correction should be made for the 27.___
   A. pressure and temperature of the aqueous phase
   B. solubility of the gas in the aqueous phase
   C. solubility of water in the gas phase
   D. vapor pressure of the water and solubility of the gas in water
   E. volume and temperature of the aqueous phase

28. The heat of combustion of C to form CO_2 is 97.600 calories per mole of carbon, that of C to form CO is 29,400 calories per mole of carbon. 28.___
Therefore, the heat of combustion of CO to CO_2 is
   A. 38,800 calories per mole
   B. 68,200 calories per mole
   C. 127,000 calories per mole
   D. 156,400 calories per mole
   E. impossible to determine from the information given

29. An intense blue solution is obtained when 29.___
   A. 0.1M sodium thiosulfate is added to 0.1M potassium permanganate
   B. 6M NH_3 is added to copper sulfide
   C. 6M NaOH is added to hydrated copper hydroxide
   D. 12M HCl is added to 1M nickel chloride solution
   E. a piece of copper is dropped into 1M $ZnSO_4$ solution

30. When 10 ml of 0.1M NaI is added to 1 ml of $Cu(NO_3)_2$, there is 30.___
   A. a blue precipitate formed
   B. a brownish solution formed
   C. a violet vapor formed
   D. a white precipitate formed
   E. no visible reaction

31. In the preparation of potassium permanganate solution, a period of storage is recommended before standardization to allow time for 31.___
   A. complete solution of the permanganate
   B. diffusion of the permanganate throughout the solution
   C. escape of air from the solution
   D. oxidation of accidentally included organic impurities, such as dust
   E. reduction of accidentally included organic impurities, such as dust

32. Comparing Solution #1 containing exactly 1 mole of NaCl and 1 mole of K_2SO_4 with Solution #2 containing 1 mole of KCl and 1 mole of Na_2SO_4, the solutions 32.___
   A. are identical
   B. differ in acidity
   C. differ in Cl^- contents
   D. differ in K^+ contents
   E. differ in $SO_4^=$ contents

33. The fats and waxes are distinguished from each other by the 33.___
  A. iodine numbers of the substances
  B. presence of distinctly different groups of fatty acids in the substances
  C. presence of glycerol in the hydrolysis products of fats and of long chain alcohols in the hydrolysis products of waxes
  D. presence of long chain alcohols in the hydrolysis products of fats but not in the hydrolysis products of waxes
  E. relative softening points of the substances

34. When hydrogen sulfide is bubbled through aqueous solutions containing mild oxidizing agents, a hazy yellow solution sometimes results. 34.___
This is caused by
  A. an insufficiency of neutral salt in the solution
  B. bubbles of undissolved gas
  C. the formation of colloidal sulfur
  D. the formation of the unstable ion $S_2O_3^{=}$
  E. the reduction of the sulfide ion

35. A neutral solution will result when equal volumes of 0.1N solutions of the following two substances are mixed: 35.___
  A. acetic acid and sodium hydroxide
  B. hydrogen chloride and ammonium hydroxide
  C. hydrogen chloride and sodium hydroxide
  D. phosphoric acid and sodium hydroxide
  E. sodium chloride and ammonium chloride

36. Cupric salts react with iodides in slightly acid solutions to give 36.___
  A. copper tri-iodide
  B. cupric iodide
  C. hydrogen ions and cuprous salts
  D. iodine and cuprous iodide
  E. a complex copper iodide ion

37. The one of the following titrations in which phenolphthalein is useful as an indicator is 37.___
  A. acetic acid with sodium hydroxide
  B. ammonia with hydrochloric acid
  C. silver nitrate with hydrochloric acid
  D. sodium acetate with hydrochloric acid
  E. sodium carbonate with hydrochloric acid

38. Of the following reagents, the one which produces a characteristic red color in the presence of nickel is 38.___
  A. Benedict's solution
  B. dimethylglyoxime
  C. 8-hydroxy quinoline
  D. magnesium perchlorate
  E. Nessler's solution

39. Potassium permanganate may be used in analytical chemistry for the direct estimation of 39.___
   A. arsenious compounds, carbonate, and ferric chloride
   B. chloride, ferrocyanide, and sulfite
   C. chloride, oxalate, and sulfate
   D. ferricyanide, oxalate, and thiosulfate
   E. ferrocyanide, oxalate, and sulfite

40. The series in which each element will displace those following it from aqueous solutions of their salts is 40.___
   A. Ag, Cu, H, Ni, Co
   B. K, Ba, Fe, Cu, Na
   C. K, Fe, H, Hg, Pt
   D. K, Zn, Pb, Pt, H
   E. Na, H, Zn, Pb, Pt

KEY (CORRECT ANSWERS)

1. A	11. C	21. B	31. D
2. D	12. E	22. A	32. D
3. A	13. D	23. C	33. C
4. C	14. A	24. B	34. C
5. C	15. B	25. A	35. C
6. B	16. E	26. E	36. D
7. D	17. A	27. D	37. A
8. A	18. A	28. B	38. B
9. A	19. A	29. C	39. E
10. B	20. C	30. D	40. C

EXAMINATION SECTION
TEST 1

DIRECTIONS: Each question or incomplete statement is followed by several suggested answers or completions. Select the one that BEST answers the question or completes the statement. *PRINT THE LETTER OF THE CORRECT ANSWER IN THE SPACE AT THE RIGHT.*

1. A compound has the empirical formula NO_2. Its molecular formula could be 1.___
   A. NO_2 B. N_2O C. N_4O_2 D. N_4O_4

2. Hydrogen bonds are STRONGEST between molecules of 2.___
   A. HBr(g) B. HI(g) C. HF(g) D. HCl(g)

3. Indicate which of the following molecule is a dipole. 3.___
   A. H_2 B. N_2 C. CH_4 D. HCl

4. Which compound is ionic? 4.___
   A. HCl B. $CaCl_2$ C. SO_2 D. N_2O

5. In the reaction $Al^{3+} + 6H_2O \rightarrow Al(H_2O)_6{}^{3+}$, the Al^{3+} ion is undergoing the process called 5.___
   A. neutralization B. addition
   C. hydrogenation D. hydration

6. Which element forms a diatomic molecule containing a triple covalent bond? 6.___
   A. Hydrogen B. Chlorine C. Nitrogen D. Oxygen

7. As the elements in Group IIA are considered from beryllium to radium, the degree of metallic activity _____ and atomic radius _____. 7.___
   A. increases; increases B. increases; decreases
   C. decreases; increases D. decreases; decreases

8. Which group of elements occur ONLY as compounds in nature because they are extremely reactive? 8.___
   A. IA B. IB C. VIA D. O

9. Which element in Period 2 has the GREATEST tendency to gain electrons? 9.___
   A. Li B. C C. F D. Ne

10. If M represents an atom of an alkali metal, the CORRECT formula for a compound of this atom with chlorine is 10.___
    A. M_2Cl B. MCl_2 C. MCl_3 D. MCl

11. At STP, 170. grams of NH_3 will occupy a total of _____ ℓ. 11.___
    A. 2.24 B. 22.4 C. 224 D. 2240

12. The total mass of iron in 1.0 mole of Fe_2O_3 is _____ grams. 12.___
A. 160 B. 112 C. 72 D. 56

13. Given the reaction: $2CO + O_2 \rightarrow 2CO_2$. 13.___
What is the MINIMUM number of moles of O_2 required to produce one mole of CO_2?
A. 1.0 B. 2.0 C. 0.25 D. 0.50

14. A solution contains 70 grams of $NaNO_3$ in 100 grams of water at 10°C. 14.___
How many additional grams of $NaNO_3$ are required to saturate this solution?
A. 10 B. 20 C. 60 D. 70

15. How do the freezing and boiling points of a sample of water change when 1 mole of NaCl is dissolved in a sample of water. 15.___
The freezing poing of the water _____, and the boiling point _____.
A. decreases; increases
B. increases; increases
C. decreases; decreases
D. increases; decreases

16. Given the reaction at equilibrium: 16.___
$H_2(g) + \frac{1}{2}O_2(g) \rightleftarrows H_2O(g) + \text{heat}$.
The value of the equilibrium constant for this reaction can be changed by
A. changing the pressure
B. changing the temperature
C. adding more O_2
D. adding a catalyst

17. Given the reaction at equilibrium: 17.___
$\frac{1}{2}N_2(g) + \frac{1}{2}O_2(g) + 21.6\ \text{kcal} \rightleftarrows NO(g)$.
The equilibrium will shift to the right if the
A. temperature increases
B. temperature decreases
C. pressure increases
D. pressure decreases

18. The heat of reaction (ΔH) is equal to the 18.___
A. heat content of the products minus the heat content of the reactants
B. heat content of the reactants minus the heat content of the products
C. entropy of the products minus the entropy of the reactants
D. entropy of the reactants minus the entropy of the products

19. Based on the table at the right, the formation of 1 mole of which of the following substances releases the GREATEST amount of energy? 19.____

A. C_2H_2
B. C_2H_4
C. $CuSO_4$
D. $BaSO_4$

Standard Energies of Formation of Compounds at 1 atm and 298 K		
Compound	Heat (Enthalpy) of Formation kcal/mole (ΔH_f°)	Free Energy of Formation kcal/mole (ΔG_f°)
Aluminum oxide Al_2O_3 (s)	−399.1	−376.8
Ammonia NH_3 (g)	−11.0	−4.0
Barium sulfate $BaSO_4$ (s)	−350.2	−323.4
Calcium hydroxide $Ca(OH)_2$ (s)	−235.8	−214.3
Carbon dioxide CO_2 (g)	−94.1	−94.3
Carbon monoxide CO (g)	−26.4	−32.8
Copper (II) sulfate $CuSO_4$ (s)	−184.0	−158.2
Ethane C_2H_6 (g)	−20.2	−7.9
Ethene C_2H_4 (g)	12.5	16.3
Ethyne (acetylene) C_2H_2 (g)	54.2	50.0
Hydrogen fluoride HF (g)	−64.2	−64.7
Hydrogen iodide HI (g)	6.2	0.3
Iodine chloride ICl (g)	4.2	−1.3
Lead (II) oxide PbO (s)	−52.4	−45.3
Magnesium oxide MgO (s)	−143.8	−136.1
Nitrogen (II) oxide NO (g)	21.6	20.7
Nitrogen (IV) oxide NO_2 (g)	8.1	12.4
Potassium chloride KCl (s)	−104.2	−97.6
Sodium chloride NaCl (s)	−98.2	−91.8
Sulfur dioxide SO_2 (g)	−71.0	−71.8
Water H_2O (g)	−57.8	−54.6
Water H_2O (ℓ)	−68.3	−56.7
Sample equation: $2Al(s) + \frac{3}{2}O_2(g) \rightarrow Al_2O_3(s)$		

20. The reaction $CH_3COOH(aq) \rightleftarrows CH_3COO^-(aq) + H^+(aq)$ has a K_a equal to 1.8×10^{-5} at 25°C. 20.____
In a solution of this acid at 25°C, the concentration of CH_3COOH is _____ the concentration of _____ ions.

A. less than; H^+
B. equal to; H^+
C. greater than; CH_3COO^-
D. equal to; CH_3COO^-

21. The conjugate base of NH_3 is 21.____

A. NH_2^-
B. NH_3
C. NO_3^-
D. NO_2^-

22. Which salt will hydrolyze in water to produce a basic solution? 22.____

A. BaI_2
B. $NaNO_2$
C. $CaCl_2$
D. $MgSO_4$

23. How many moles of KOH are needed to exactly neutralize 500. ml of a 1.0 M HCl solution? 23.____

A. 1.0
B. 2.0
C. 0.25
D. 0.50

24. In a 0.01 M solution of HCl, litmus will be _____ and phenolphthalein will be _____. 24.____

A. blue; colorless
B. blue; pink
C. red; colorless
D. red; pink

25. Pure water is similar to 0.1 M HCl in that they both 25.____

A. contain H_3O^+ ions
B. are neutral to litmus
C. are good conductors of electricity
D. have a pH greater than 7

26. Which half-reaction CORRECTLY represents reduction? 26.___
A. $Sn^{2+} + 2e^- \rightarrow Sn^{4+}$
B. $Sn^{2+} \rightarrow Sn^{4+} + 2e^-$
C. $Sn^{2+} + 2e^- \rightarrow Sn^0$
D. $Sn^{2+} \rightarrow Sn^0 + 2e^-$

27. Which ion is MOST easily oxidized? 27.___
A. Br^- B. Cl^- C. F^- D. I^-

28. In the reaction $MnO_2 + 4HCl \rightarrow MnCl_2 + 2H_2O + Cl_2$, which species is reduced? 28.___
A. Mn^{4+} B. O^{2-} C. H^+ D. Cl^-

29. During the electrolysis of fused NaCl, which half-reaction occurs at the negative electrode? 29.___
A. $Na^+ + 1e^- \rightarrow Na^0$
B. $Na^0 \rightarrow Na^+ + 1e^-$
C. $2Cl^- \rightarrow Cl_2^0 + 2e^-$
D. $Cl_2^0 + 2e^- \rightarrow 2Cl^-$

30. Given the reaction: 30.___
$\_Cr^{3+} + 10\ OH^- + \_ClO_3^- \rightarrow \_CrO_4^{2-} + \_Cl^- + 5H_2O$.
When the reaction is completely balanced using the smallest whole numbers, the coefficient of Cr^{3+} will be
A. 1 B. 2 C. 3 D. 4

31. An organic acid is represented by the formula 31.___
A. CH_3COOH B. CH_3OH C. CH_3OCH_3 D. CH_3COOCH_3

32. Which equation represents an esterification reaction? 32.___
A. $C_6H_{12}O_6 \rightarrow 2C_2H_5OH + 2CO_2$
B. $C_5H_{10} + H_2 \rightarrow C_5H_{12}$
C. $C_3H_8 + Cl_2 \rightarrow C_3H_7Cl + HCl$
D. $HCOOH + CH_3OH \rightarrow HCOOCH_3 + HOH$

33. The compound $CH_3CH_2CH_2CH_3$ belongs to the series that has the general formula 33.___
A. C_nH_{2n-2} B. C_nH_{2n+2} C. C_nH_{n-6} D. C_nH_{n+6}

34. Which molecule contains a triple covalent bond between adjacent carbon atoms? 34.___
A. C_2H_4 B. C_2H_2 C. C_3H_6 D. C_3H_8

35. The total number of OH groups in a molecule of glycerol is 35.___
A. 1 B. 2 C. 3 D. 4

KEY (CORRECT ANSWERS)

1. A	11. C	21. A	31. A
2. C	12. B	22. B	32. D
3. D	13. D	23. D	33. B
4. B	14. A	24. C	34. B
5. D	15. A	25. A	35. C
6. C	16. B	26. C	
7. A	17. A	27. D	
8. A	18. A	28. A	
9. C	19. D	29. A	
10. D	20. C	30. B	

TEST 2

DIRECTIONS: Each question or incomplete statement is followed by several suggested answers or completions. Select the one that BEST answers the question or completes the statement. *PRINT THE LETTER OF THE CORRECT ANSWER IN THE SPACE AT THE RIGHT.*

1. According to the chart at the right, a compound which forms spontaneously from its elements is 1.____
   A. NO
   B. NO_2
   C. C_2H_4
   D. C_2H_6

Standard Energies of Formation of Compounds at 1 atm and 298 K

Compound	*Heat (Enthalpy) of Formation kcal/mole* (ΔH_f°)	*Free Energy of Formation kcal/mole* (ΔG_f°)
Aluminum oxide Al_2O_3 (s)	−399.1	−376.8
Ammonia NH_3 (g)	−11.0	−4.0
Barium sulfate $BaSO_4$ (s)	−350.2	−323.4
Calcium hydroxide $Ca(OH)_2$ (s)	−235.8	−214.3
Carbon dioxide CO_2 (g)	−94.1	−94.3
Carbon monoxide CO (g)	−26.4	−32.8
Copper (II) sulfate $CuSO_4$ (s)	−184.0	−158.2
Ethane C_2H_6 (g)	−20.2	−7.9
Ethene C_2H_4 (g)	12.5	16.3
Ethyne (acetylene) C_2H_2 (g)	54.2	50.0
Hydrogen fluoride HF (g)	−64.2	−64.7
Hydrogen iodide HI (g)	6.2	0.3
Iodine chloride ICl (g)	4.2	−1.3
Lead (II) oxide PbO (s)	−52.4	−45.3
Magnesium oxide MgO (s)	−143.8	−136.1
Nitrogen (II) oxide NO (g)	21.6	20.7
Nitrogen (IV) oxide NO_2 (g)	8.1	12.4
Potassium chloride KCl (s)	−104.2	−97.6
Sodium chloride NaCl (s)	−98.2	−91.8
Sulfur dioxide SO_2 (g)	−71.0	−71.8
Water H_2O (g)	−57.8	−54.6
Water H_2O (ℓ)	−68.3	−56.7

Sample equation

$2Al\ (s) + \frac{3}{2}O_2\ (g) \rightarrow Al_2O_3\ (s)$

2. A neutralization reaction is demonstrated by the equation 2.____
   A. $NaOH + HCl \rightarrow NaCl + H_2O$
   B. $2Na + 2H_2O \rightarrow 2NaOH + H_2$
   C. $Zn + CuSO_4 \rightarrow ZnSO_4 + Cu$
   D. $AgNO_3 + NaCl \rightarrow AgCl + NaNO_3$

3. The concentration of H_3O^+ ions, in moles per liter, of a 0.0001 M HCl solution is 1 × 3.____
   A. 10^{-1} B. 10^{-2} C. 10^{-3} D. 10^{-4}

4. As HF dissolves in water, the following ionization reaction occurs: $HF + H_2O \rightleftarrows H_3O^+ + F^-$. 4.____
   In this reaction, a proton is donated to
   A. H_3O^+ by H_2O B. H_2O by HF
   C. H_3O^+ by F^- D. HF by F^-

5. How many milliliters of 1.0 M H_2SO_4 are needed to EXACTLY neutralize 15 milliliters of 2.0 M $Ba(OH)_2$? 5.____
   ______ ml.
   A. 7.5 B. 10. C. 15 D. 30.

6. A 0.1 M solution of which acid is the BEST conductor of electricity at 25°C? 6.____
   A. $H_3PO_4 (K_a = 7.1 \times 10^{-3})$
   B. $HNO_2 (K_a = 5.1 \times 10^{-4})$
   C. $CH_3COOH (K_a = 1.8 \times 10^{-5})$
   D. $H_2S (K_a = 1.0 \times 10^{-7})$

7. What is the pH of a solution if the hydroxide ion concentration is 1×10^{-7} mole per liter? 7.____
   A. 1　　B. 7　　C. 10　　D. 14

8. Which is a redox reaction? 8.____
   A. $Mg + 2HCl \rightarrow MgCl_2 + H_2$
   B. $Mg(OH)_2 + 2HCl \rightarrow MgCl_2 + 2H_2O$
   C. $Mg^{2+}(aq) + 2OH^-(aq) \rightarrow Mg(OH)_2$
   D. $MgCl_2 + 6H_2O \rightarrow MgCl_2 \cdot 6H_2O$

9. In the reaction $Zn + Cu^{2+} \rightarrow Zn^{2+} + Cu$, the oxidizing agent 9.____
   A. gains protons　　B. loses electrons
   C. is reduced　　D. is oxidized

10. A chemical cell has a net reaction of $Cu + 2Ag^+ = Cu^{2+} + 2Ag$. 10.____
At equilibrium, the cell potential, in volts, is
   A. -0.46　　B. 0.00　　C. +0.34　　D. +0.80

Questions 11-12.

DIRECTIONS: Questions 11 and 12 are to be answered on the basis of the following chart.

STANDARD ELECTRODE POTENTIALS

Ionic Concentrations 1 M Water at 298 K, 1 atm

Half-Reaction	$E°$ *(volts)*
$F_2(g) + 2e^- \longrightarrow 2F^-$	+2.87
$MnO_4^- + 8H^+ + 5e^- \longrightarrow Mn^{2+} + 4H_2O$	+1.52
$Au^{3+} + 3e^- \longrightarrow Au(s)$	+1.50
$Cl_2(g) + 2e^- \longrightarrow 2Cl^-$	+1.36
$Cr_2O_7^{2-} + 14H^+ + 6e^- \longrightarrow 2Cr^{3+} + 7H_2O$	+1.33
$MnO_2(s) + 4H^+ + 2e^- \longrightarrow Mn^{2+} + 2H_2O$	+1.28
$\frac{1}{2}O_2(g) + 2H^+ + 2e^- \longrightarrow H_2O$	+1.23
$Br_2(\ell) + 2e^- \longrightarrow 2Br^-$	+1.06
$NO_3^- + 4H^+ + 3e^- \longrightarrow NO(g) + 2H_2O$	+0.96
$\frac{1}{2}O_2(g) + 2H^+(10^{-7}M) + 2e^- \longrightarrow H_2O$	+0.82
$Ag^+ + e^- \longrightarrow Ag(s)$	+0.80
$\frac{1}{2}Hg_2^{2+} + e^- \longrightarrow Hg(\ell)$	+0.79
$Hg^{2+} + 2e^- \longrightarrow Hg(\ell)$	+0.78
$NO_3^- + 2H^+ + e^- \longrightarrow NO_2(g) + H_2O$	+0.78
$Fe^{3+} + e^- \longrightarrow Fe^{2+}$	+0.77
$I_2(s) + 2e^- \longrightarrow 2I^-$	+0.53
$Cu^+ + e^- \longrightarrow Cu(s)$	+0.52
$Cu^{2+} + 2e^- \longrightarrow Cu(s)$	+0.34
$SO_4^{2-} + 4H^+ + 2e^- \longrightarrow SO_2(g) + 2H_2O$	+0.17
$Sn^{4+} + 2e^- \longrightarrow Sn^{2+}$	+0.15
$2H^+ + 2e^- \longrightarrow H_2(g)$	0.00
$Pb^{2+} + 2e^- \longrightarrow Pb(s)$	−0.13
$Sn^{2+} + 2e^- \longrightarrow Sn(s)$	−0.14
$Ni^{2+} + 2e^- \longrightarrow Ni(s)$	−0.25
$Co^{2+} + 2e^- \longrightarrow Co(s)$	−0.28
$2H^+(10^{-7}M) + 2e^- \longrightarrow H_2(g)$	−0.41
$Fe^{2+} + 2e^- \longrightarrow Fe(s)$	−0.44
$Cr^{3+} + 3e^- \longrightarrow Cr(s)$	−0.74
$Zn^{2+} + 2e^- \longrightarrow Zn(s)$	−0.76
$2H_2O + 2e^- \longrightarrow 2OH^- + H_2(g)$	−0.83
$Mn^{2+} + 2e^- \longrightarrow Mn(s)$	−1.18
$Al^{3+} + 3e^- \longrightarrow Al(s)$	−1.66
$Mg^{2+} + 2e^- \longrightarrow Mg(s)$	−2.37
$Na^+ + e^- \longrightarrow Na(s)$	−2.71
$Ca^{2+} + 2e^- \longrightarrow Ca(s)$	−2.87
$Sr^{2+} + 2e^- \longrightarrow Sr(s)$	−2.89
$Ba^{2+} + 2e^- \longrightarrow Ba(s)$	−2.90
$Cs^+ + e^- \longrightarrow Cs(s)$	−2.92
$K^+ + e^- \longrightarrow K(s)$	−2.92
$Rb^+ + e^- \longrightarrow Rb(s)$	−2.93
$Li^+ + e^- \longrightarrow Li(s)$	−3.00

11. Which half-reaction has a reduction potential (E°) of +1.50 volts? 11.___

A. $Au^{3+} + 3e^- \rightarrow Au(s)$ B. $Cr^{3+} + 3e^- \rightarrow Cr(s)$
C. $Sn^{2+} + 2e^- \rightarrow Sn(s)$ D. $Ba^{2+} + 2e^- \rightarrow Ba(s)$

12. Which metal will react spontaneously with H^+? 12.___

A. Au   B. Ag   C. Cr   D. Cu

13. Given the reaction: $3Mg(s) + 2Cr^{3+}(aq) \rightarrow 3Mg^{2+}(aq) + 2Cr(s)$ 13.___
What is the potential (E°) for the overall reaction?
_____ volts.

A. +3.11   B. -3.11   C. +1.63   D. -1.63

14. In the reaction $4NH_3 + 5O_2 \rightarrow 4NO + 6H_2O$, the oxidation number of nitrogen changes from 14.___

A. -2 to -3   B. -2 to +3   C. -3 to -2   D. -3 to +2

15. An unsaturated hydrocarbon is represented by the formula 15.___

A. C_3H_8 B. C_3H_7Cl C. C_3H_6 D. CCl_4

16. The isomers 1-chloropropane and 2-chloropropane differ ONLY in 16.___

A. molecular composition
B. molecular structure
C. the number of chloro groups per molecule
D. the number of carbon atoms per molecule

17. Which structural formula represents ethene? 17.___

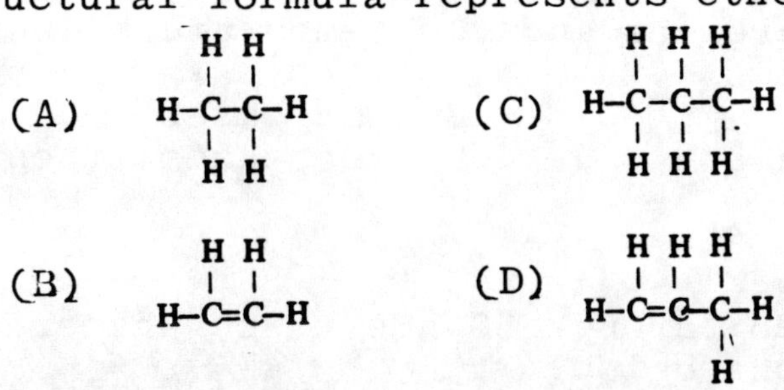

18. Which diagram may be used to represent a benzene ring? 18.___

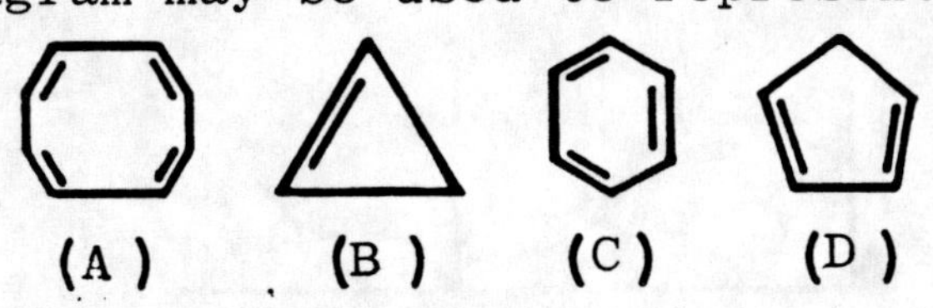

19. A compound with the structural formula

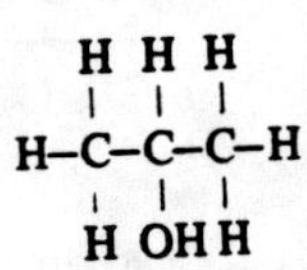

is classified as an 19.___

A. alcohol
B. acid
C. alkane
D. alkene

20. As the temperature of a liquid *increases*, its vapor pressure 20.____
A. decreases
B. increases
C. remains the same
D. increases, then decreases

21. As a sulfur atom gains electrons, its radius 21.____
A. decreases
B. increases
C. remains the same
D. increases, then decreases

22. As the H_3O^+ ion concentration of a solution increases and the OH^- concentration decreases, the pH of the solution 22.____
A. decreases
B. increases
C. remains the same
D. cannot be determined from the information given

23. Given the system $CO_2(s) \rightleftarrows CO_2(g)$ at equilibrium. As the pressure increases at constant temperature, the amount of $CO_2(g)$ will 23.____
A. decrease
B. increase
C. remain the same
D. decrease, then increase

24. When a catalyst lowers the activation energy of a reaction, the rate of the reaction 24.____
A. decreases
B. increases
C. remains the same
D. decreases, then increases

25. A sample of a gas is at STP. As the pressure decreases and the temperature increases, the volume of the gas 25.____
A. decreases
B. increases
C. remains the same
D. decreases, then increases

Questions 26-30.

DIRECTIONS: Questions 26 through 30 are to be answered on the basis of the graphs below.

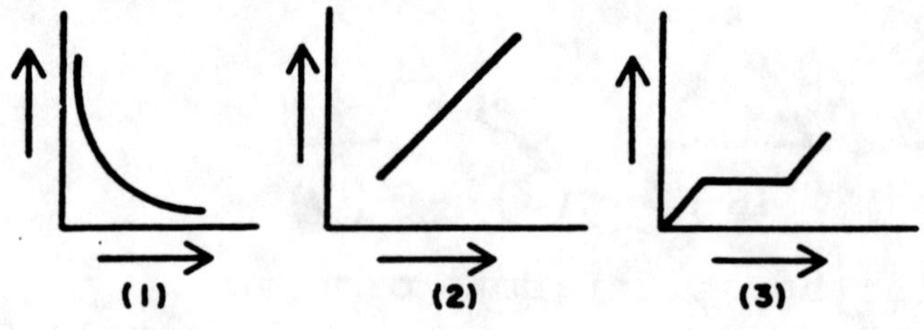

26. Which graph BEST represents how the volume of a given mass of a gas varies with the Kelvin (absolute) temperature at constant pressure? 26.____
A. 1
B. 2
C. 3
D. None of the above

27. Which graph BEST represents how the volume of a given mass of a gas varies with the pressure exerted on it at constant temperature? 27.____
   A. 1
   B. 2
   C. 3
   D. None of the above

28. Which represents a homogeneous mixture? 28.____
   A. $CuSO_4(s)$ B. $Br_2(I)$ C. $NaCl(aq)$ D. $CO_2(g)$

29. A 16.0 gram sample of $CH_4(g)$ is at 0°C and 1 atmosphere. The volume of the gas sample in liters at 27°C and 1 atmosphere is equal to 29.____
   A. $16.0 \times \frac{1}{27}$
   B. $16.0 \times \frac{27}{1}$
   C. $22.4 \times \frac{273}{300}$
   D. $22.4 \times \frac{300}{273}$

30. At STP, 32 grams of O_2 would occupy the same volume as 30.____
   A. 64 g of H_2
   B. 32 g of SO_2
   C. 8.0 g of CH_4
   D. 4.0 g of He

31. The mass number of a $^{3}_{1}H$ atom is 31.____
   A. 1 B. 2 C. 3 D. 4

32. Which orbital notation CORRECTLY represents the outermost principal energy level of a sulfur atom in the ground state? 32.____

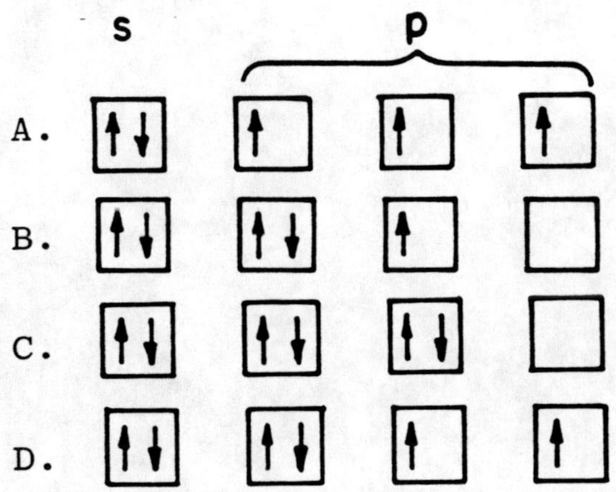

33. A radioactive source emits radiation which is deflected as shown in the diagram at the right. 33.____
This radiation could be
   A. $_{-1}^{0}e$
   B. $^{4}_{2}He$
   C. $^{1}_{1}H$
   D. $^{1}_{0}n$

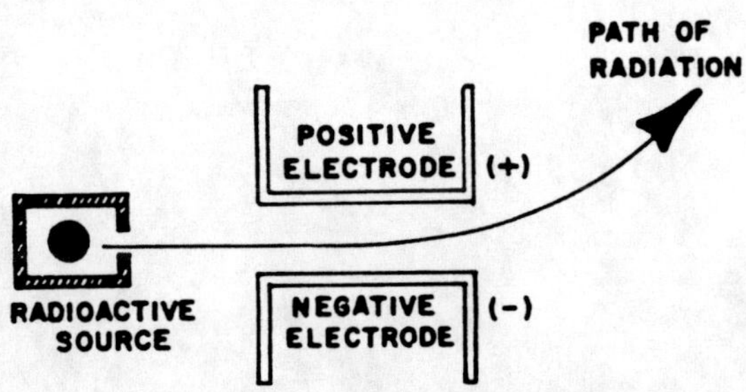

34. In an aluminum atom in the ground state, the energy level which contains the MOST electrons has the principal quantum number 34.___
A. 1 B. 2 C. 3 D. 4

35. A fluoride ion (F^-) has the same electron configuration as 35.___
A. Na B. Na^+ C. Cl D. Cl^-

KEY (CORRECT ANSWERS)

1. D	11. A	21. B	31. C
2. A	12. C	22. A	32. D
3. D	13. C	23. A	33. A
4. B	14. D	24. B	34. B
5. D	15. C	25. B	35. B
6. A	16. B	26. B	
7. B	17. B	27. A	
8. A	18. C	28. C	
9. C	19. A	29. D	
10. B	20. B	30. D	

TEST 3

DIRECTIONS: Each question or incomplete statement is followed by several suggested answers or completions. Select the one that BEST answers the question or completes the statement. *PRINT THE LETTER OF THE CORRECT ANSWER IN THE SPACE AT THE RIGHT.*

1. The molecules of ______ have the HIGHEST average kinetic energy. 1.___
   A. He(g) at 0°C B. CO_2(g) at 20°C
   C. HCl(g) at 40°C D. N_2(g) at 60°C

2. The temperature of 50 grams of water was raised to 50°C by the addition of 1,000 calories of heat energy. The initial temperature of the water was _____ °C. 2.___
   A. 10 B. 20 C. 30 D. 60

3. Which gas will MOST closely resemble an ideal gas at STP? 3.___
   A. SO_2 B. NH_3 C. Cl_2 D. H_2

4. When water freezes, each gram loses an amount of heat equal to its heat of 4.___
   A. fusion B. vaporization
   C. sublimation D. reaction

5. Equal volumes of all gases at the same temperature and pressure contain an equal number of 5.___
   A. molecules B. atoms C. electrons D. protons

6. Which is the electron configuration of an atom in the ground state? 6.___
   A. $1s^2 2s^1 2p^2$ B. $1s^2 2s^2 2p^5 3s^2$
   C. $1s^2 2s^2 3s^1$ D. $1s^2 2s^2 2p^6 3s^1$

7. An electron in an atom will emit energy when it moves from energy level 7.___
   A. 2s to 3p B. 2s to 2p C. 2p to 3s D. 2p to 1s

8. Which atom has a COMPLETELY filled 3rd principal energy level? 8.___
   A. Ar B. Zn C. Ca D. K

9. The electron dot symbol $:\ddot{\underset{..}{X}}:^-$ represents an ion of atom X. Atom X could be an atom of 9.___
   A. K B. H C. I D. S

10. Which of the following elements has the HIGHEST first ionization energy? 10.___
    A. Li B. Na C. K D. Rb

11. There are no known stable isotopes of the element with an atomic number of 11.___
A. 20 B. 38 C. 56 D. 88

12. The TOTAL number of protons found in an OH^- ion is 12.___
A. 1 B. 8 C. 9 D. 17

13. Two atoms of element A unite to form a molecule with the formula A_2. 13.___
The bond between the atoms in the molecule is
A. electrovalent B. ionic
C. nonpolar covalent D. polar covalent

14. Which bond has the GREATEST degree of ionic character? 14.___
A. H-Cl B. I-Cl C. Cl-Cl D. K-Cl

15. Which substance will conduct electricity in both the solid phase and the liquid phase? 15.___
A. AgCl B. Ag C. H_2 D. HCl

16. Indicate which of the following statements BEST explains why a CH_4 molecule is nonpolar. 16.___
A. C and H are nonmetals.
B. C and H have the same electronegativity.
C. CH_4 has a symmetrical charge distribution.
D. CH_4 is a gas at room temperature.

17. Mercury (I) chloride is CORRECTLY represented by the formula 17.___
A. HG_2Cl B. $HgCl_2$ C. Hg_2Cl_2 D. Hg_2Cl_4

18. An example of an empirical formula is 18.___
A. C_2H_2 B. H_2O_2 C. C_2Cl_2 D. $CaCl_2$

19. When the equation $H_2 + N_2 \rightarrow NH_3$ is completely balanced using smallest whole numbers, the sum of all the coefficients will be 19.___
A. 6 B. 7 C. 3 D. 12

20. Which particle has the LARGEST radius? 20.___
A. Cu B. Cu^{2+} C. Se D. Se^{2-}

21. Which atom may form a negative ion with the electron configuration $1s^2$? 21.___
A. H B. He C. Li D. Be

22. Group _____ is known as the halogens. 22.___
A. IA B. IIA C. VIIA D. O

23. Which element exhibits BOTH metallic and nonmetallic properties? 23.___
A. B B. Ba C. K D. Kr

24. In which group are all of the elements solids at STP? 24.____
A. VIIA B. VIA C. VA D. IVA

25. A white anhydrous powder that dissolves in water to form a blue aqueous solution could be 25.____
A. $MgSO_4$ B. $BaSO_4$ C. $CuSO_4$ D. $CaSO_4$

26. There are ______ $\times 10^{23}$ molecules in 0.25 mole of O_2. 26.____
A. 12 B. 6.0 C. 3.0 D. 1.5

27. A 10.0 gram sample of a hydrate was heated until all the water of hydration was driven off. The mass of anhydrous product remaining was 8.00 grams. 27.____
The percent of water in the hydrate is ______ percent.
A. 12.5 B. 20.0 C. 25.0 D. 80.0

28. A compound contains 50% sulfur and 50% oxygen by mass. What is the empirical formula of the compound? 28.____
A. SO B. SO_2 C. SO_3 D. SO_4

29. The density, in grams per liter, of N_2 gas at STP is 29.____
A. 28.0 B. 14.0 C. 1.25 D. 0.800

30. Given the balanced equation: $3Fe + 4H_2O \rightarrow Fe_3O_4 + 4H_2$. When 36.0 grams of H_2O is consumed, the TOTAL number of liters of H_2 produced at STP is 30.____
A. 22.4 B. 33.6 C. 44.8 D. 89.6

31. What is the TOTAL number of grams of KCl (formula mass = 74.6) in 1.00 liter of 0.200 molar solution? 31.____
______ grams.
A. 7.46 B. 14.9 C. 22.4 D. 29.8

32. According to the table at the right, in which reaction do the products have a HIGHER energy content than the reactants? 32.____

Heats of Reaction at 1 atm and 298K

Reaction	ΔH (kcal)
$CH_4(g) + 2O_2(g) \rightarrow CO_2(g) + 2H_2O(\ell)$	—212.8
$C_3H_8(g) + 5O_2(g) \rightarrow 3CO_2(g) + 4H_2O(\ell)$	—530.6
$CH_3OH(\ell) + \frac{3}{2}O_2(g) \rightarrow CO_2(g) + 2H_2O(\ell)$	—173.6
$C_6H_{12}O_6(s) + 6O_2(g) \rightarrow 6CO_2(g) + 6H_2O(\ell)$	—669.9
$CO(g) + \frac{1}{2}O_2(g) \rightarrow CO_2(g)$	—67.7
$NaOH(s) \xrightarrow{H_2O} Na^+(aq) + OH^-(aq)$	—10.6
$NH_4Cl(s) \xrightarrow{H_2O} NH_4^+(aq) + Cl^-(aq)$	+3.5
$H^+(aq) + OH^-(aq) \rightarrow H_2O(\ell)$	—13.8

A. $CH_4(g) + 2O_2(g) \rightarrow CO_2(g) + 2H_2O(\ell)$

B. $CH_3OH(\ell) + \frac{3}{2}O_2(g) \rightarrow CO_2(g) + 2H_2O(\ell)$

C. $NH_4Cl(s) \xrightarrow{H_2O} NH_4^+(aq) + Cl^-(aq)$

D. $NaOH(s) \xrightarrow{H_2O} Na^+(aq) + OH^-(aq)$

33. According to the table at the right, which compound is MOST soluble in water? 33.____

A. $BaCO_3$
B. $BaSO_4$
C. $ZnCO_3$
D. $ZnSO_4$

TABLE OF SOLUBILITIES IN WATER

i — nearly insoluble
ss — slightly soluble
s — soluble
d — decomposes
n — not isolated

	acetate	bromide	carbonate	chloride	chromate	hydroxide	iodide	nitrate	phosphate	sulfate	sulfide
Aluminum	ss	s	n	s	n	i	s	s	i	s	d
Ammonium	s	s	s	s	s	s	s	s	s	s	s
Barium	s	s	i	s	i	s	s	s	i	i	d
Calcium	s	s	i	s	s	ss	s	s	i	ss	d
Copper II	s	s	i	s	i	i	d	s	i	s	i
Iron II	s	s	i	s	n	i	s	s	i	s	i
Iron III	s	s	n	s	i	i	s	s	i	ss	d
Lead	s	ss	i	ss	i	i	ss	s	i	i	i
Magnesium	s	s	i	s	s	i	s	s	i	s	d
Mercury I	ss	i	i	i	ss	n	i	s	i	ss	i
Mercury II	s	ss	i	s	ss	i	i	s	i	d	i
Potassium	s	s	s	s	s	s	s	s	s	s	s
Silver	ss	i	i	i	ss	n	i	s	i	ss	i
Sodium	s	s	s	s	s	s	s	s	s	s	s
Zinc	s	s	i	s	s	i	s	s	i	s	i

34. Given the reaction $H_2(g) + I_2(g) \rightleftarrows 2HI(g)$. What is the CORRECT equilibrium expression for this reaction? 34.____

$K_{eq} =$

A. $\frac{[HI]^2}{[H_2][I_2]}$ B. $\frac{[HI]^2}{[2H][2I]}$ C. $\frac{[H_2][I_2]}{[HI]}$ D. $\frac{[HI]^2}{[H]^2[I]^2}$

35. Given the reaction at equilibrium: $A(g) + B(g) \rightleftarrows AB(g)$. Indicate which equilibrium constant, K_{eq}, MOST favors the formation of AB(g)? 35.____

A. 1×10^{-3} B. 2×10^{-6} C. 3×10^{-9} D. 4×10^{-12}

KEY (CORRECT ANSWERS)

1. D
2. C
3. D
4. A
5. A

6. D
7. D
8. B
9. C
10. A

11. D
12. C
13. C
14. D
15. B

16. C
17. C
18. D
19. A
20. D

21. A
22. C
23. A
24. D
25. C

26. D
27. B
28. B
29. C
30. C

31. B
32. C
33. D
34. A
35. A

EXAMINATION SECTION

TEST 1

DIRECTIONS: Each question or incomplete statement is followed by several suggested answers or completions. Select the one that BEST answers the question or completes the statement. *PRINT THE LETTER OF THE CORRECT ANSWER IN THE SPACE AT THE RIGHT.*

1. The reactivity of the metals in Groups IA and IIA generally INCREASES with 1.___
   A. *increased* ionization energy
   B. *increased* atomic radius
   C. *decreased* nuclear charge
   D. *decreased* mass

2. The heat of vaporization for water is 540. calories per gram. 2.___
   What is the MINIMUM number of calories needed to change 40.0 grams of water at 100°C to steam at the same temperature and pressure?
   A. 43,200 B. 21,600 C. 540. D. 40.0

3. A compound contains 40% calcium, 12% carbon, and 48% oxygen by mass. 3.___
   What is the empirical formula of this compound?
   A. $CaCO_3$ B. CaC_2O_4 C. CaC_3O_6 D. $CaCO_2$

4. If the density of a gas at STP is 2.50 grams per liter, what is the gram molecular mass of the gas? 4.___
   A. 2.50 B. 22.4 C. 56.0 D. 89.6

5. Which solution will freeze at the LOWEST temperature? 5.___
   _____ of sugar in _____ grams of water.
   A. 1 mole; 500 B. 1 mole; 1,000
   C. 2 moles; 500 D. 2 moles; 1,000

6. What is the total number of carbon atoms contained in 22.4 liters of CO gas at STP? 6.___
   A. 1.00 B. 0.500
   C. 3.01×10^{23} D. 6.02×10^{23}

7. Which statement is TRUE if the free energy (ΔG) of a reaction is zero? 7.___
   The
   A. rate of the forward reaction is zero
   B. rate of the reverse reaction is zero
   C. reaction is approaching equilibrium
   D. reaction is at equilibrium

8. According to the chart at the right, which compound is MORE soluble than $BaSO_4$ at 1 atmosphere and 298 K? 8.____

A. AgBr
B. $PbCl_2$
C. AgI
D. ZnS

Constants for Various Equilibria at 1 atm and 298 K	
$H_2O = H^+(aq) + OH^-(aq)$	$K_w = 1.0 \times 10^{-14}$
$CH_3COO^-(aq) + H_2O = CH_3COOH(aq) + OH^-(aq)$	$K_b = 5.6 \times 10^{-10}$
$NH_3(aq) + H_2O = NH_4^+(aq) + OH^-(aq)$	$K_b = 1.8 \times 10^{-5}$
$CO_3^{2-}(aq) + H_2O = HCO_3^-(aq) + OH^-(aq)$	$K_b = 2.1 \times 10^{-4}$
$Ag(NH_3)_2^+(aq) = Ag^+(aq) + 2NH_3(aq)$	$K_{eq} = 6.3 \times 10^{-8}$
$N_2(g) + 3H_2(g) = 2NH_3(g)$	$K_{eq} = 6.7 \times 10^{5}$
$H_2(g) + I_2(g) = 2HI(g)$	$K_{eq} = 3.5 \times 10^{-1}$

Compound	K_{sp}	*Compound*	K_{sp}
AgCl	1.6×10^{-10}	$PbCl_2$	1.6×10^{-5}
AgBr	7.7×10^{-13}	$PbCrO_4$	1.8×10^{-14}
AgI	1.5×10^{-16}	PbI_2	1.4×10^{-8}
$BaSO_4$	1.1×10^{-10}	ZnS	1.6×10^{-23}

9. A chemical reaction will ALWAYS occur spontaneously if the reaction has a 9.____

A. negative ΔG
B. positive ΔG
C. negative ΔH
D. positive ΔH

10. The free energy change, ΔG, must be negative when ΔH is ______ and ΔS is ______. 10.____

A. positive; positive
B. positive; negative
C. negative; positive
D. negative; negative

11. Given the reaction at equilibrium: 11.____

$NaCl(s) = Na^+(aq) + Cl^-(aq)$.

The addition of KCl to this system will cause a shift in the equilibrium to the _____ and the concentration of $Na^+(aq)$ ions will _____.

A. left; increase
B. right; increase
C. left; decrease
D. right; decrease

12. What is the pH of an aqueous solution of $C_6H_{12}O_6$? 12.____

A. 1 B. 7 C. 11 D. 14

13. A sample of pure water contains 13.____

A. neither OH^- ions nor H_3O^+ ions
B. equal concentrations of OH^- and H_3O^+ ions
C. a larger concentration of H_3O^+ ions than OH^- ions
D. a smaller concentration of H_3O^+ ions than OH^- ions

14. Which of the following acids ionizes to the LEAST extent at 298 K? 14.____

A. HF B. HNO_2 C. H_2S D. H_2O

15. According to the chart at the right, which substance is amphiprotic? 15.___

A. H_2SO_4
B. HNO_3
C. NH_3
D. HBr

RELATIVE STRENGTHS OF ACIDS IN AQUEOUS SOLUTION at 1 atm AND 298 K

Conjugate Pairs		
ACID	BASE	K_a
$HI \rightarrow$	$H^+ + I^-$	very large
$HBr \rightarrow$	$H^+ + Br^-$	very large
$HCl \rightarrow$	$H^+ + Cl^-$	very large
$HNO_3 \rightarrow$	$H^+ + NO_3^-$	very large
$H_2SO_4 \rightarrow$	$H^+ + HSO_4^-$	large
$H_2O + SO_2 \rightarrow$	$H^+ + HSO_3^-$	1.7×10^{-2}
$HSO_4^- \rightarrow$	$H^+ + SO_4^{2-}$	1.3×10^{-2}
$H_3PO_4 \rightarrow$	$H^+ + H_2PO_4^-$	7.1×10^{-3}
$Fe(H_2O)_6^{3+} \rightarrow$	$H^+ + Fe(H_2O)_5(OH)^{2+}$	6.0×10^{-3}
$HF \rightarrow$	$H^+ + F^-$	6.7×10^{-4}
$HNO_2 \rightarrow$	$H^+ + NO_2^-$	5.1×10^{-4}
$Cr(H_2O)_6^{3+} \rightarrow$	$H^+ + Cr(H_2O)_5(OH)^{2+}$	1.0×10^{-4}
$CH_3COOH \rightarrow$	$H^+ + CH_3COO^-$	1.8×10^{-5}
$Al(H_2O)_6^{3+} \rightarrow$	$H^+ + Al(H_2O)_5(OH)^{2+}$	1.0×10^{-5}
$H_2O + CO_2 \rightarrow$	$H^+ + HCO_3^-$	4.4×10^{-7}
$H_2S \rightarrow$	$H^+ + HS^-$	1.0×10^{-7}
$H_2PO_4^- \rightarrow$	$H^+ + HPO_4^{2-}$	6.3×10^{-8}
$HSO_3^- \rightarrow$	$H^+ + SO_3^{2-}$	6.2×10^{-8}
$NH_4^+ \rightarrow$	$H^+ + NH_3$	5.7×10^{-10}
$HCO_3^- \rightarrow$	$H^+ + CO_3^{2-}$	4.7×10^{-11}
$HPO_4^{2-} \rightarrow$	$H^+ + PO_4^{3-}$	4.4×10^{-13}
$HS^- \rightarrow$	$H^+ + S^{2-}$	1.3×10^{-14}
$OH^- \rightarrow$	$H^+ + O^{2-}$	$< 10^{-36}$
$NH_3 \rightarrow$	$H^+ + NH_2^-$	very small

16. As 0.1 M HCl is added to 0.1 M KOH, the pH of the basic solution ______ and basicity ______. 16.___

A. decreases; decreases
B. increases; decreases
C. decreases; increases
D. increases; increases

17. In order for a redox reaction to be spontaneous, the potential (E°) for the overall reaction must be 17.___

A. greater than zero
B. zero
C. between zero and -1
D. less than -1

18. Given the reaction: 18.___

$2Au^{3+}(aq) + 3Ni^0 \rightarrow 2Au^0 + 3Ni^{2+}(aq)$.

The cell potential (E°) for the overall reaction is ______ volts.

A. 3.75
B. 2.25
C. 1.75
D. 1.25

19. Which half-reaction occurs at the cathode in an electrolytic cell in which an object is being plated with copper? 19.___

A. $Cu(s) \rightarrow Cu^{2+} + 2e^-$
B. $Cu(s) + 2e^- \rightarrow Cu^{2+}$
C. $Cu^{2+} \rightarrow Cu(s) + 2e^-$
D. $Cu^{2+} + 2e^- \rightarrow Cu(s)$

20. Which will oxidize Zn(s) to Zn^{2+} but will NOT oxidize Pb(s) to Pb^{2+}? 20.___

A. Al^{3+}
B. Au^{3+}
C. Co^{2+}
D. Mg^{2+}

21. According to the table at the right, which halogen will react spontaneously with Au(s) to produce Au^{3+}? 21.____

A. Br_2
B. F_2
C. I_2
D. Cl_2

STANDARD ELECTRODE POTENTIALS

Ionic Concentrations 1 M Water at 298 K, 1 atm

Half-Reaction	$E°$ *(volts)*
$F_2(g) + 2e^- \rightarrow 2F^-$	+2.87
$MnO_4^- + 8H^+ + 5e^- \rightarrow Mn^{2+} + 4H_2O$	+1.52
$Au^{3+} + 3e^- \rightarrow Au(s)$	+1.50
$Cl_2(g) + 2e^- \rightarrow 2Cl^-$	+1.36
$Cr_2O_7^{2-} + 14H^+ + 6e^- \rightarrow 2Cr^{3+} + 7H_2O$	+1.33
$MnO_2(s) + 4H^+ + 2e^- \rightarrow Mn^{2+} + 2H_2O$	+1.28
$\frac{1}{2}O_2(g) + 2H^+ + 2e^- \rightarrow H_2O$	+1.23
$Br_2(\ell) + 2e^- \rightarrow 2Br^-$	+1.06
$NO_3^- + 4H^+ + 3e^- \rightarrow NO(g) + 2H_2O$	+0.96
$\frac{1}{2}O_2(g) + 2H^+(10^{-7}M) + 2e^- \rightarrow H_2O$	+0.82
$Ag^+ + e^- \rightarrow Ag(s)$	+0.80
$\frac{1}{2}Hg_2^{2+} + e^- \rightarrow Hg(\ell)$	+0.79
$Hg^{2+} + 2e^- \rightarrow Hg(\ell)$	+0.78
$NO_3^- + 2H^+ + e^- \rightarrow NO_2(g) + H_2O$	+0.78
$Fe^{3+} + e^- \rightarrow Fe^{2+}$	+0.77
$I_2(s) + 2e^- \rightarrow 2I^-$	+0.53
$Cu^+ + e^- \rightarrow Cu(s)$	+0.52
$Cu^{2+} + 2e^- \rightarrow Cu(s)$	+0.34
$SO_4^{2-} + 4H^+ + 2e^- \rightarrow SO_2(g) + 2H_2O$	+0.17
$Sn^{4+} + 2e^- \rightarrow Sn^{2+}$	+0.15
$2H^+ + 2e^- \rightarrow H_2(g)$	0.00
$Pb^{2+} + 2e^- \rightarrow Pb(s)$	−0.13
$Sn^{2+} + 2e^- \rightarrow Sn(s)$	−0.14
$Ni^{2+} + 2e^- \rightarrow Ni(s)$	−0.25
$Co^{2+} + 2e^- \rightarrow Co(s)$	−0.28
$2H^+(10^{-7}M) + 2e^- \rightarrow H_2(g)$	−0.41
$Fe^{2+} + 2e^- \rightarrow Fe(s)$	−0.44
$Cr^{3+} + 3e^- \rightarrow Cr(s)$	−0.74
$Zn^{2+} + 2e^- \rightarrow Zn(s)$	−0.76
$2H_2O + 2e^- \rightarrow 2OH^- + H_2(g)$	−0.83
$Mn^{2+} + 2e^- \rightarrow Mn(s)$	−1.18
$Al^{3+} + 3e^- \rightarrow Al(s)$	−1.66
$Mg^{2+} + 2e^- \rightarrow Mg(s)$	−2.37
$Na^+ + e^- \rightarrow Na(s)$	−2.71
$Ca^{2+} + 2e^- \rightarrow Ca(s)$	−2.87
$Sr^{2+} + 2e^- \rightarrow Sr(s)$	−2.89
$Ba^{2+} + 2e^- \rightarrow Ba(s)$	−2.90
$Cs^+ + e^- \rightarrow Cs(s)$	−2.92
$K^+ + e^- \rightarrow K(s)$	−2.92
$Rb^+ + e^- \rightarrow Rb(s)$	−2.93
$Li^+ + e^- \rightarrow Li(s)$	−3.00

22. Which is the structure for 1,2-dibromoethane? 22.____

A. 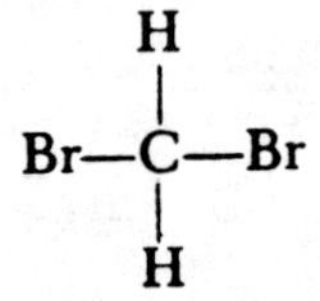

C. 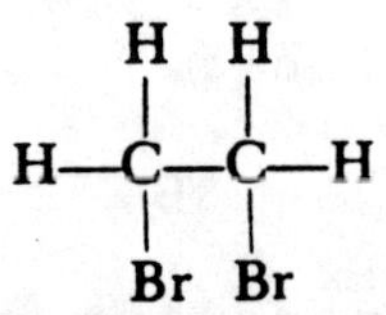

B. 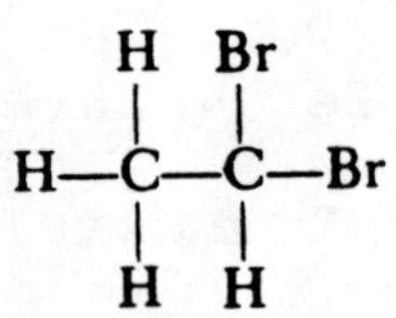

D. 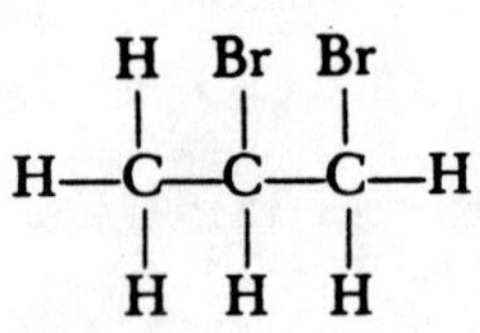

23. Which organic reaction involves the bonding of monomers by a dehydration process? 23.____

A. Substitution
B. Oxidation
C. Addition polymerization
D. Condensation polymerization

24. Which is the CORRECT structural formula for glycerol? 24.___

A. 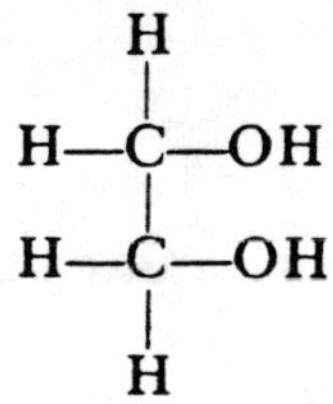

B. 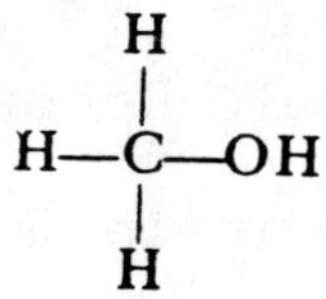

C. 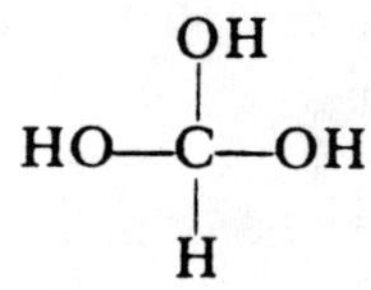

D. 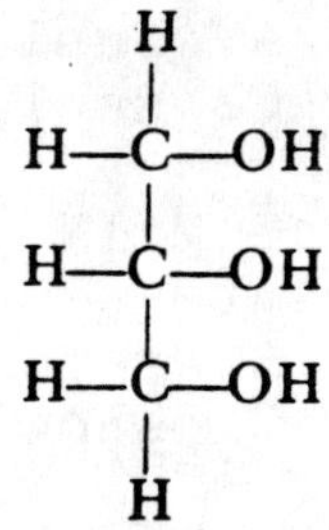

25. Which is represented by the structural formula shown at the right? 25.___

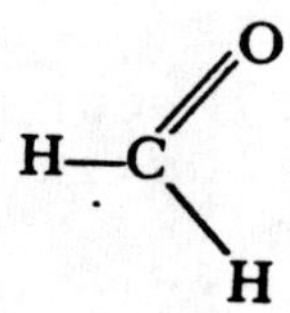

An
- A. aldehyde
- B. alcohol
- C. alkane
- D. acid

26. Which is the general formula for an ether? 26.___

A. $R—OH$

B. 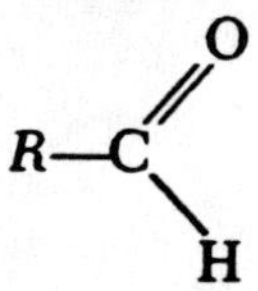

C. $R_1—O—R_2$

D. 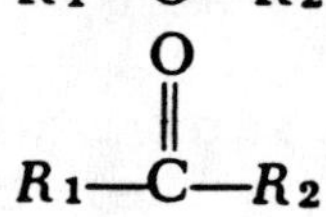

27. In crude petroleum, fractions can be separated according to their differing boiling points by 27.___
- A. the contact process
- B. the Haber process
- C. fractional distillation
- D. cracking

28. Which type of chemical reaction is the corrosion of iron? 28.___
- A. Reduction-oxidation
- B. Substitution
- C. Polymerization
- D. Decomposition

29. Which equation represents a simple example of cracking? 29.___
- A. $N_2 + 3H_2 \xrightarrow{600°C} 2NH_3$
- B. $S + O_2 \rightarrow SO_2$
- C. $C_3H_8 + 5O_2 \rightarrow 3CO_2 + 4H_2O$
- D. $C_{14}H_{30} \xrightarrow{600°C} C_7H_{16} + C_7H_{14}$

30. The Haber process is used to produce 30.___
- A. sulfur dioxide
- B. ammonia
- C. sulfuric acid
- D. sodium chloride

31. Which of the following metals forms a self-protective coating when exposed to air and moisture? 31.___
- A. Zinc
- B. Calcium
- C. Iron
- D. Sodium

32. When a uranium nucleus breaks up into fragments, which type of nuclear reaction occurs? 32.___

A. Fusion
B. Fission
C. Replacement
D. Redox

33. The MAIN purpose of the moderator in a fission reactor is to 33.___

A. remove heat
B. produce heat
C. slow down neutrons
D. speed up neutrons

34. In the reaction $X + {}^{1}_{1}H \rightarrow {}^{6}_{3}Li + {}^{4}_{2}He$, the nucleus represented by X is 34.___

A. ${}^{9}_{3}Li$
B. ${}^{10}_{5}B$
C. ${}^{9}_{4}Be$
D. ${}^{10}_{6}C$

35. Which particle can NOT be accelerated by the electric or magnetic fields in a particle accelerator? 35.___

A. Neutron
B. Proton
C. Alpha particle
D. Beta particle

36. Which equation represents artificial transmutation? 36.___

A. $H_2O \rightarrow H^+ + OH^-$
B. $UF_6 + 6Na \rightarrow 6NaF + U$
C. ${}^{238}_{92}U \rightarrow {}^{234}_{90}Th + {}^{4}_{2}He$
D. ${}^{27}_{13}Al + {}^{4}_{2}He \rightarrow {}^{30}_{15}P + {}^{1}_{0}n$

37. The diagram at the right shows a eudiometer tube in a water bath. The water levels inside and outside the tube are equal. If the atmospheric pressure is 750 torr, the pressure inside the tube is _____ torr. 37.___

A. 740
B. 750
C. 760
D. 800

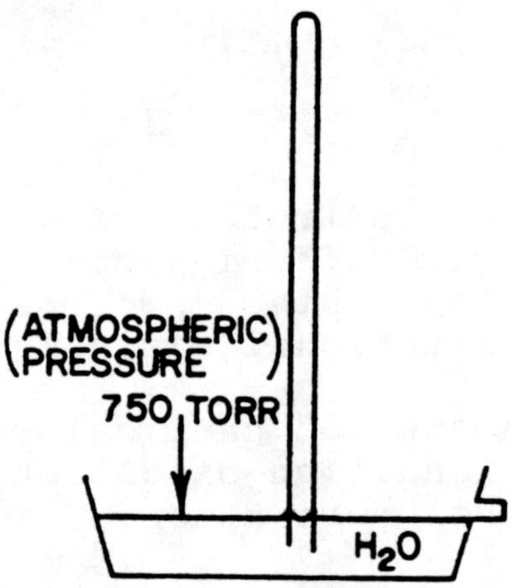

38. The volume of an acid required to neutralize exactly 15.00 milliliters (ml) of a base could be measured MOST precisely if it were added to the base solution from a 38.___

A. 100 ml graduate
B. 125 ml Erlenmeyer flask
C. 50 ml buret
D. 50 ml beaker

Questions 39-40.

DIRECTIONS: Questions 39 and 40 are to be answered on the basis of the table below which shows the data collected during the heating of a 5.0 gram sample of a hydrated salt.

Mass of Salt (g)	Heating Time (min)
5.0	0.0
4.1	5.0
3.1	10.
3.0	15.
3.0	30.
3.0	60.

39. After 60 minutes, how many grams of water appear to remain in the salt? 39.___
A. 0.00
B. 2.0
C. 1.9
D. 0.90

40. The original sample contains _____ percent water. 40.___
A. 82.
B. 60.
C. 30.
D. 40.

KEY (CORRECT ANSWERS)

1. D	11. C	21. B	31. A
2. B	12. B	22. B	32. B
3. A	13. B	23. D	33. C
4. C	14. D	24. D	34. C
5. C	15. C	25. A	35. A
6. D	16. A	26. C	36. D
7. D	17. A	27. C	37. B
8. B	18. C	28. A	38. C
9. A	19. D	29. D	39. A
10. C	20. C	30. B	40. D

TEST 2

DIRECTIONS: Each question or incomplete statement is followed by several suggested answers or completions. Select the one that BEST answers the question or completes the statement. *PRINT THE LETTER OF THE CORRECT ANSWER IN THE SPACE AT THE RIGHT.*

1. Given the system at chemical equilibrium: 1.____
   $2O_3(g) \rightleftharpoons 3O_2(g)$ ($K_{eq} = 2.5 \times 10^{12}$).

   The concentration of O_3 and O_2 must be
   A. constant B. equal
   C. increasing D. decreasing

2. If HCl and H_2O react together in an acid-base reaction to form their Brönsted-Lowry conjugates, the products would be 2.____
   A. HCl and H_3O^+ B. Cl^- and OH^-
   C. Cl_2 and H_2 D. Cl^- and H_3O^+

3. What is the hydronium ion concentration of a solution at 298 K whose hydroxide ion concentration is 1×10^{-8}? 3.____
   1 ×
   A. 10^{-6} B. 10^{-7} C. 10^{-8} D. 10^{-14}

4. A 0.1 M acid solution at 298 K would conduct electricity BEST if the acid had a K_a value of 4.____
   A. 1.0×10^{-7} B. 1.8×10^{-5}
   C. 6.7×10^{-4} D. 1.7×10^{-2}

5. If 50 milliliters (ml) of a 0.01 M HCl solution is required to neutralize exactly 25 milliliters (ml) of NaOH, what is the concentration of the base? 5.____
   ______ M.
   A. 0.01 B. 0.02 C. 0.0005 D. 0.04

6. When tested, a solution turns red litmus to blue. This indicates that the solution contains more 6.____
   A. H^+ ions than OH^- ions
   B. H_3O^+ ions than OH^- ions
   C. OH^- ions than H_3O^+ ions
   D. H^+ and OH^- ions than H_2O molecules

7. Oxygen has a positive oxidation number in the compound 7.____
   A. H_2O B. H_2O_2 C. OF_2 D. IO_2

8. In a chemical cell composed of two half-cells, ions are allowed to flow from one half-cell to another by means of 8.____
   A. electrodes B. an external conductor
   C. a voltmeter D. a salt bridge

9. Which occurs in the half-reaction $Na(s) \rightarrow Na^+ + e^-$? 9.___
   A. Na(s) is reduced. B. Na(s) is oxidized.
   C. Na(s) gains electrons. D. Na^+ is oxidized.

10. What is the oxidizing agent in the reaction 10.___
   $Zn^0 + 2Ag^+ \rightarrow Zn^{2+} + 2Ag^0$?
   A. Zn^0 B. Ag^0 C. Zn^{2+} D. Ag^+

11. Which is a redox reaction? 11.___
   A. $CaCO_3 \rightarrow CaO + CO_2$
   B. $NaOH + HCl \rightarrow NaCl + H_2O$
   C. $2NH_4Cl + Ca(OH)_2 \rightarrow 2NH_3 + 2H_2O + CaCl_2$
   D. $2H_2O \rightarrow 2H_2 + O_2$

12. Given the unbalanced equation: $Ca^0 + Al^{3+} \rightarrow Ca^{2+} + Al^0$. 12.___
   When the equation is completely balanced with the SMALLEST whole-number coefficients, what is the coefficient of Ca^0?
   A. 1 B. 2 C. 3 D. 4

13. Organic compounds that are essentially non-polar and exhibit weak intermolecular forces have 13.___
   A. low melting points
   B. low vapor pressure
   C. high conductivity in solution
   D. high boiling points

14. What is the formula for pentanol? 14.___
   A. C_5H_{12} B. $C_5H_{11}OH$ C. C_4H_{10} D. C_4H_9OH

15. The functional group 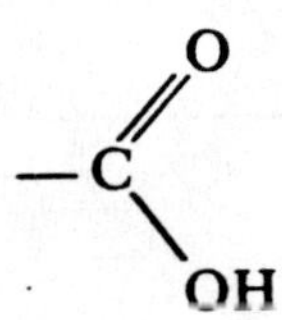 is ALWAYS found in an organic 15.___

   A. acid
   B. ester
   C. ether
   D. aldehyde

16. Which compound is a member of the alkene series of hydrocarbons? 16.___
   A. Benzene B. Propene C. Toluene D. Butadiene

17. Which compound contains a triple bond? 17.___
   A. CH_4 B. C_2H_2 C. C_3H_6 D. C_4H_{10}

18. As additional KCl is added to a saturated solution of KCl, the conductivity of the solution 18.___
   A. decreases
   B. increases
   C. remains the same
   D. cannot be determined from the information given

19. As the pressure on a sample of gas increases at constant temperature, the volume of the gas 19.___
   A. decreases
   B. increases
   C. remains the same
   D. cannot be determined from the information given

20. As 1 gram of $H_2O(\ell)$ changes to 1 gram of $H_2O(s)$, the entropy of the system 20.____
   A. decreases
   B. increases
   C. remains the same
   D. cannot be determined from the information given

21. Given the reaction at equilibrium: 21.____
   $A(g) + B(g) + heat = AB(g)$.
   As the pressure increases at constant temperature, the value of the equilibrium constant
   A. decreases
   B. increases
   C. remains the same
   D. cannot be determined from the information given

22. The graph at the right represents changes of state for an unknown substance. What is the boiling temperature of the substance? 22.____
   ______°C.
   A. 0
   B. 20
   C. 70
   D. 40

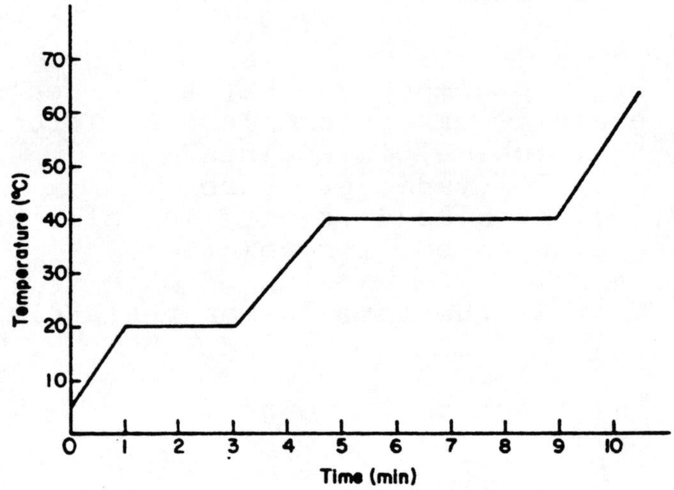

23. Which is the formula of a binary compound? 23.____
   A. KOH B. $NaClO_3$ C. Al_2S_3 D. $Bi(NO_3)_3$

24. At constant pressure, 50. milliliters (ml) of a gas at 20.°C is heated to 30.°C. 24.____
   The new volume of the gas in milliliters (ml) is equal to 50. ×
   A. $\frac{20.}{30.}$ B. $\frac{30.}{20.}$ C. $\frac{293}{303}$ D. $\frac{303}{293}$

25. Which of the following substances is made up of particles with the HIGHEST average kinetic energy? 25.____
   ______°C.
   A. $Fe(s)$ at 35 B. $Br_2(\ell)$ at 20
   C. $H_2O(\ell)$ at 30 D. $CO_2(g)$ at 25

26. Solid substances are MOST likely to sublime if they have _____ vapor pressure and _____ intermolecular attractions. 26.___
A. high; strong
B. high; weak
C. low; strong
D. low; weak

27. Which of the following atoms has the GREATEST nuclear charge? 27.___
A. $^{14}_{7}N$
B. $^{12}_{6}C$
C. $^{2}_{1}H$
D. $^{4}_{2}He$

28. Potassium-42 requires _____ hours to undergo three half-life periods. 28.___
A. 6.2
B. 12.4
C. 24.8
D. 37.2

29. Usually, the term *kernel* includes all parts of the atom EXCEPT the 29.___
A. neutrons
B. protons
C. valence electrons
D. orbital electrons

30. Which nuclear emission moving through an electric field would be deflected toward the positive electrode? 30.___
A. Alpha particle
B. Beta particle
C. Gamma radiation
D. Proton

31. What is the electron configuration for Be^{2+} ions? 31.___
A. $1s^1$
B. $1s^2$
C. $1s^2 2s^1$
D. $1s^2 2s^2$

32. What type of bonding is found in the molecule HBr? 32.___
A. Ionic
B. Metallic
C. Nonpolar covalent
D. Polar covalent

33. What is the name of the calcium salt of sulfuric acid? Calcium 33.___
A. thiosulfate
B. sulfate
C. sulfide
D. sulfite

34. What is the formula of nitrogen (I) oxide? 34.___
A. NO
B. N_2O
C. NO_2
D. N_2O_4

35. When a reaction occurs between atoms with ground state electron configurations $1s^2 2s^1$ and $1s^2 2s^2 2p^5$, the predominant type of bond formed is 35.___
A. polar covalent
B. nonpolar covalent
C. ionic
D. metallic

36. Which represents both an empirical and a molecular formula? 36.___
A. P_2O_5
B. N_2O_4
C. C_3H_6
D. $C_6H_{12}O_6$

37. In Period 2, as the elements are considered from left to right, there is a DECREASE in 37.___
A. ionization energy
B. atomic mass
C. metallic character
D. nonmetallic character

38. Which molecule is relatively inactive and contains a triple bond? 38.___
A. N_2 B. O_2 C. Cl_2 D. H_2

39. Atoms of metallic elements tend to _____ electrons and form _____ ions. 39.___
A. gain; negative B. gain; positive
C. lose; negative D. lose; positive

40. Alkali metals, alkaline earth metals, and halogens are elements found respectively in Groups 40.___
A. IA, IIA, and O B. IIA, IIIA, and VIIA
C. IA, IIA, and IVA D. IA, IIA, and VIIA

KEY (CORRECT ANSWERS)

1. A	11. D	21. C	31. B
2. D	12. C	22. D	32. D
3. A	13. A	23. C	33. B
4. D	14. B	24. D	34. B
5. B	15. A	25. A	35. C
6. C	16. B	26. B	36. A
7. C	17. B	27. A	37. C
8. D	18. C	28. D	38. A
9. B	19. A	29. C	39. D
10. D	20. A	30. B	40. D

TEST 3

DIRECTIONS: Each question or incomplete statement is followed by several suggested answers or completions. Select the one that BEST answers the question or completes the statement. *PRINT THE LETTER OF THE CORRECT ANSWER IN THE SPACE AT THE RIGHT.*

1. A 1-liter flask contains two gases at a total pressure of 3.0 atmospheres. If the partial pressure of one of the gases is 0.5 atmosphere, then the partial pressure of the other gas MUST be _____ atm. 1.___
   A. 1.0 B. 2.5 C. 1.5 D. 0.50

2. Which of the following BEST describes exothermic chemical reactions? 2.___
   They
   A. *never* release heat
   B. *always* release heat
   C. *never* occur spontaneously
   D. *always* occur spontaneously

3. What is the TOTAL number of molecules in 1.0 mole of Cl_2(ag)? 3.___
   A. 35 B. 70
   C. 6.0×10^{23} D. 12×10^{24}

4. The number of calories needed to raise the temperature of 10 grams of water from 20°C to 30°C is 4.___
   A. 10 B. 20 C. 100 D. 40

5. The heat of fusion for ice is 80 calories per gram. Adding 80 calories of heat to one gram of ice at STP will cause the ice to _____ temperature. 5.___
   A. increase in
   B. decrease in
   C. change to water at a higher
   D. change to water at the same

6. Which 1-mole sample of atoms requires the LEAST energy to form a mole of positive ions? 6.___
   A. Ge B. Ca C. Ga D. K

7. The atomic number of an atom is ALWAYS equal to the total number of 7.___
   A. neutrons in the nucleus
   B. protons in the nucleus
   C. neutrons plus protons in the atom
   C. protons plus electrons in the atom

8. Which electron configuration represents an atom in an excited state? 8.___
   A. $1s^2 2s^2 2p^6 3p^1$ B. $1s^2 2s^2 2p^6 3s^2 3p^1$
   C. $1s^2 2s^2 2p^6 3s^2 3p^2$ D. $1s^2 2s^2 2p^6 3s^2$

9. Which principal energy level can hold a MAXIMUM of 18 electrons? 9.___
   A. 5 B. 2 C. 3 D. 4

10. Which is the CORRECT electron dot representation of an atom of sulfur in the ground state? 10.___

   A. S: B. ·Ṡ: C. ·S̤̈: D. :S̤̈:

11. The TOTAL number of d orbitals in the third principal energy level is 11.___
   A. one B. five C. three D. seven

12. Which of the following nuclei is an isotope of (10p 11n) ? 12.___

   A. (10p 9n) B. (11p 10n) C. (9p 11n) D. (11p 12n)

13. When the equation $NH_3 + O_2 \rightarrow HNO_3 + H_2O$ is completely balanced using SMALLEST whole numbers, the coefficient of O_2 would be 13.___
   A. 1 B. 2 C. 3 D. 4

14. The bonding in NH_3 is MOST similar to the bonding in 14.___
   A. H_2O B. NaCl C. MgO D. KF

15. Which is the formula of an ionic compound? 15.___
   A. SO_2 B. CO_2 C. CH_3OH D. NaOH

16. Which electron dot formula represents a molecule that contains a nonpolar covalent bond? 16.___

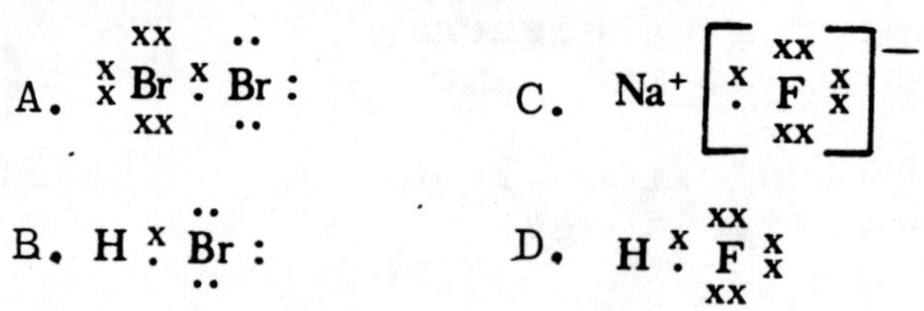

17. Molecule-ion attractions are found in 17.___
   A. Cu(s) B. CO(g) C. KBr(ℓ) D. NaCl(aq)

18. Hydrogen bonds are STRONGEST between the molecules of 18.___
   A. HF(ℓ) B. HCl(ℓ) C. HBr(ℓ) D. HI(ℓ)

19. The bonds present in silicon carbide (SiC) are 19.___
   A. covalent B. ionic
   C. metallic D. van der Waals

20. As the elements are considered from the top to the bottom of Group VA, which sequence in properties occurs? 20.___
   A. Metal → metalloid → nonmetal
   B. Metal → nonmetal → metalloid
   C. Metalloid → metal → nonmetal
   D. Nonmetal → metalloid → metal

21. The element found in Group IIIA and in Period 2 is 21.___
   A. Be B. Mg C. B D. Al

22. Which element is considered malleable? 22.___
   A. Gold B. Hydrogen C. Sulfur D. Radon

23. The MOST active nonmetal in the Periodic Table of the Elements is 23.___
   A. Na B. F C. I D. Cl

24. Element M has an electronegativity of less than 1.2 and reacts with bromine to form the compound MBr_2. 24.___
Element M could be
   A. Al B. Na C. Ca D. K

25. A chloride dissolves in water to form a colored solution. The chloride could be 25.___
   A. HCl B. KCl C. $CaCl_2$ D. $CuCl_2$

26. Which of the following particles has the SMALLEST radius? 26.___
   A. Na^0 B. K^0 C. Na^+ D. K^+

27. The gram molecular mass of CO_2 is the same as the gram molecular mass of 27.___
   A. CO B. SO_2 C. C_2H_6 D. C_3H_8

28. The empirical formula of a compound is CH_2 and its molecular mass is 70. 28.___
What is the molecular formula of the compound?
   A. C_2H_2 B. C_2H_4 C. C_4H_{10} D. C_5H_{10}

29. Consider the following reaction: 29.___
   $2Na + 2H_2O \rightarrow 2NaOH + H_2$.
What is the TOTAL number of moles of hydrogen produced when 4 moles of sodium react completely?
   A. One B. Two C. Three D. Four

30. The percent by mass of nitrogen in $Mg(CN)_2$ is equal to _____ × 100. 30.___
   A. $\frac{14}{76}$ B. $\frac{14}{50}$ C. $\frac{28}{76}$ D. $\frac{28}{50}$

31. What is the molarity of a solution of KNO_3 (molecular mass = 101) that contains 404 grams of KNO_3 in 2.00 liters of solution? 31.___
   A. 1.00 B. 2.00 C. 0.500 D. 4.00

32. The graph at the right represents the potential energy changes that occur in a chemical reaction. Which letter represents the activated complex? 32.___

A. A
B. B
C. C
D. D

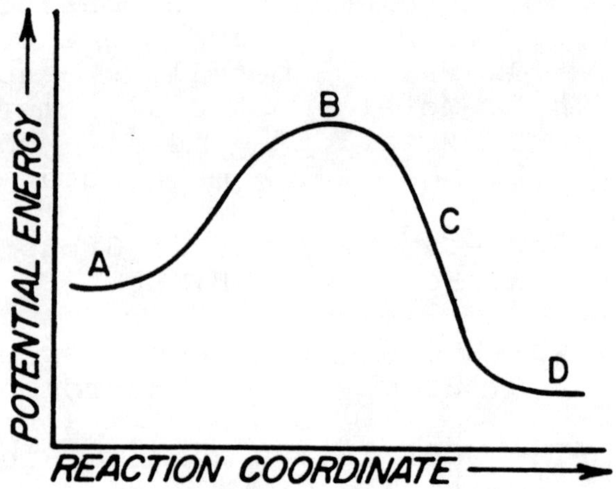

33. According to the chart at the right, which of the following substances is MOST soluble at 60°C? 33.___

A. NH_4Cl
B. KCl
C. NaCl
D. NH_3

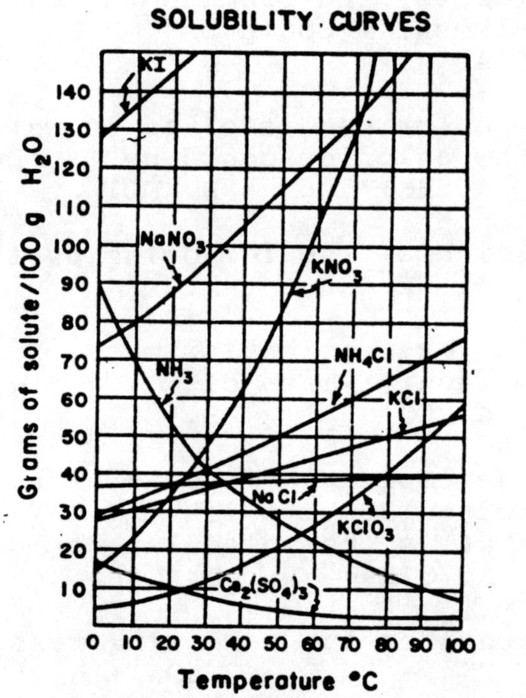

34. In a chemical reaction, the difference in potential energy between the products and the reactants is equal to 34.___

A. ΔS
B. ΔG
C. ΔH
D. ΔT

35. Given the reaction at equilibrium: 35.___
$2CO(g) + O_2(g) = 2CO_2(g)$.
The CORRECT equilibrium expression for this reaction is
K_{eq} =

A. $\frac{[2CO][O_2]}{[2CO_2]}$ B. $\frac{[CO]^2[O_2]}{[CO_2]^2}$

C. $\frac{[2CO_2]}{[2CO][O_2]}$ D. $\frac{[CO_2]^2}{[CO]^2[O_2]}$

KEY (CORRECT ANSWERS)

1. B	11. B	21. C	31. B
2. B	12. A	22. A	32. B
3. C	13. B	23. B	33. A
4. C	14. A	24. C	34. C
5. D	15. D	25. D	35. D
6. D	16. A	26. C	
7. B	17. D	27. D	
8. A	18. A	28. D	
9. C	19. A	29. B	
10. C	20. D	30. C	

EXAMINATION SECTION

TEST 1

DIRECTIONS: Each question or incomplete statement is followed by several suggested answers or completions. Select the one that BEST answers the question or completes the statement. *PRINT THE LETTER OF THE CORRECT ANSWER IN THE SPACE AT THE RIGHT.*

1. Which substance is an ionic solid? 1.___
   A. LiCl B. HCl C. Ne D. Fe

2. The element in Group VIA whose isotopes are all radioactive is 2.___
   A. O B. S C. Te D. Po

3. Which particle has the LARGEST radius? 3.___
   A. F^- B. Ne C. Cl^- D. Ar

4. Which element occurs in nature ONLY in compounds? 4.___
   A. Au B. Ag C. Ne D. Na

5. In the Periodic Table, the transition elements are found in the 5.___
   A. A groups *only*
   B. B groups *only*
   C. A groups and Group O
   D. B groups and Group VIII

6. In the ground state, how many electrons are in the outermost s sublevel of each element in Group VIIA? 6.___
   A. Five B. Two C. Seven D. Eight

7. The gram-molecular mass of a gas is 44.0 grams. Its density at STP is _____ g/ℓ. 7.___
   A. 0.509 B. 1.43 C. 1.96 D. 2.84

8. At standard pressure, a 1 molal solution of sugar has a boiling point _____ than 100°C and a freezing point than 0°C. 8.___
   A. greater; greater
   B. greater; less
   C. less; greater
   D. less; less

9. In the reaction $N_2 + 3H_2 \rightarrow 2NH_3$, _____ gram(s) of H_2 are needed to produce EXACTLY 1 mole of ammonia. 9.___
   A. one B. two C. three D. four

10. The volume of 16.0 grams of $SO_2(g)$ at STP is CLOSEST to _____ ℓ. 10.___
    A. 5.60 B. 11.2 C. 22.4 D. 44.8

11. A compound is 92.3% carbon and 7.7% hydrogen. The empirical formula of this compound is 11.___
    A. CH B. CH_2 C. CH_3 D. CH_4

12. The ΔG of a chemical reaction refers to the change in 12.____
A. entropy
B. state
C. activation energy
D. free energy

13. Based on the chart at the right, a compound which forms spontaneously from its elements is 13.____
A. ethene
B. ethane
C. nitrogen (II) oxide
D. nitrogen (IV) oxide

Standard Energies of Formation of Compounds at 1 atm and 298 K

Compound	Heat (Enthalpy) of Formation kcal/mole (ΔH_f°)	Free Energy of Formation kcal/mole (ΔG_f°)
Aluminum oxide Al_2O_3 (s)	—399.1	—376.8
Ammonia NH_3 (g)	—11.0	—4.0
Barium sulfate $BaSO_4$ (s)	—350.2	—323.4
Calcium hydroxide $Ca(OH)_2$ (s)	—235.8	—214.3
Carbon dioxide CO_2 (g)	—94.1	—94.3
Carbon monoxide CO (g)	—26.4	—32.8
Copper (II) sulfate $CuSO_4$ (s)	—184.0	—158.2
Ethane C_2H_6 (g)	—20.2	—7.9
Ethene C_2H_4 (g)	12.5	16.3
Ethyne (acetylene) C_2H_2 (g)	54.2	50.0
Hydrogen fluoride HF (g)	—64.2	—64.7
Hydrogen iodide HI (g)	6.2	0.3
Iodine chloride ICl (g)	4.2	—1.3
Lead (II) oxide PbO (s)	—52.4	—45.3
Magnesium oxide MgO (s)	—143.8	—136.1
Nitrogen (II) oxide NO (g)	21.6	20.7
Nitrogen (IV) oxide NO_2 (g)	8.1	12.4
Potassium chloride KCl (s)	—104.2	—97.6
Sodium chloride NaCl (s)	—98.2	—91.8
Sulfur dioxide SO_2 (g)	—71.0	—71.8
Water H_2O (g)	—57.8	—54.6
Water H_2O (ℓ)	—68.3	—56.7

Sample equation

$2Al(s) + \frac{3}{2}O_2(g) \rightarrow Al_2O_3(s)$

14. Given the reaction at equilibrium: 14.____
$AgI(s) \rightleftarrows Ag^+(aq) + I^-(aq)$.
Which is the CORRECT solubility product constant expression for the reaction?
K_{sp} =
A. $[Ag^+][I^-]$
B. $\frac{1}{[Ag^+][I^-]}$
C. $[Ag^+] + [I^-]$
D. $\frac{1}{[Ag^+] + [I^-]}$

15. Given the reaction at equilibrium: 15.____
$CH_3COOH(aq) + H_2O(\ell) \rightleftarrows H_3O^+(aq) + CH_3COO^-(aq)$.
The addition of which ion will cause an INCREASE in the concentration of $CH_3COO^-(aq)$?
A. H_3O^+
B. Cl^-
C. Na^+
D. OH^-

16. Given the reaction at equilibrium: 16.____
$H_2(g) + Cl_2(g) \rightleftarrows 2HCl(g)$ + heat.
The equilibrium will shift to the right when there is an INCREASE in
A. temperature
B. pressure
C. concentration of $H_2(g)$
D. concentration of $HCl(g)$

17. How many milliliters of 5.0 M NaOH are needed to EXACTLY neutralize 40. milliliters of 2.0 M HCl? 17.____

A. 8.0 B. 10. C. 16 D. 40.

18. A 0.1 M HCl solution DIFFERS from a 0.1 M NaOH solution in that the HCl solution 18.____

A. has a lower pH
B. turns litmus blue
C. contains H_3O^+ ions
D. does not contain OH^-

19. Which compound is a strong electrolyte? 19.____

A. CH_3COOH B. H_2S C. $C_{12}H_{22}O_{11}$ D. HNO_3

20. What is the $[H_3O^+]$ of a 0.001 M NaOH solution? 20.____

1 ×

A. 10^{-1} B. 10^{-7} C. 10^{-11} D. 10^{-14}

21. What are the two Brönsted acids in the reaction below? 21.____

$HPO_4^{2-} + H_2O = PO_4^{3-} + H_3O^+$

A. HPO_4^{2-} and PO_4^{3-}
B. HPO_4^{2-} and H_3O^+
C. H_2O and H_3O^+
D. H_2O and PO_4^{3-}

22. In an electrolytic cell, a Cl^- ion would be attracted to the _____ electrode and _____. 22.____

A. positive; oxidized
B. positive; reduced
C. negative; oxidized
D. negative; reduced

23. Based on the chart at the right, which pair will react spontaneously? 23.____

A. I_2 and F^-
B. I_2 and Cl^-
C. I_2 and Ag
D. I_2 and Zn

STANDARD ELECTRODE POTENTIALS

Ionic Concentrations 1 M Water at 298 K, 1 atm

Half-Reaction	$E°$ *(volts)*
$F_2(g) + 2e^- \rightarrow 2F^-$	+2.87
$MnO_4^- + 8H^+ + 5e^- \rightarrow Mn^{2+} + 4H_2O$	+1.52
$Au^{3+} + 3e^- \rightarrow Au(s)$	+1.50
$Cl_2(g) + 2e^- \rightarrow 2Cl^-$	+1.36
$Cr_2O_7^{2-} + 14H^+ + 6e^- \rightarrow 2Cr^{3+} + 7H_2O$	+1.33
$MnO_2(s) + 4H^+ + 2e^- \rightarrow Mn^{2+} + 2H_2O$	+1.28
$\frac{1}{2}O_2(g) + 2H^+ + 2e^- \rightarrow H_2O$	+1.23
$Br_2(\ell) + 2e^- \rightarrow 2Br^-$	+1.06
$NO_3^- + 4H^+ + 3e^- \rightarrow NO(g) + 2H_2O$	+0.96
$\frac{1}{2}O_2(g) + 2H^+(10^{-7}M) + 2e^- \rightarrow H_2O$	+0.82
$Ag^+ + e^- \rightarrow Ag(s)$	+0.80
$\frac{1}{2}Hg_2^{2+} + e^- \rightarrow Hg(\ell)$	+0.79
$Hg^{2+} + 2e^- \rightarrow Hg(\ell)$	+0.78
$NO_3^- + 2H^+ + e^- \rightarrow NO_2(g) + H_2O$	+0.78
$Fe^{3+} + e^- \rightarrow Fe^{2+}$	+0.77
$I_2(s) + 2e^- \rightarrow 2I^-$	+0.53
$Cu^+ + e^- \rightarrow Cu(s)$	+0.52
$Cu^{2+} + 2e^- \rightarrow Cu(s)$	+0.34
$SO_4^{2-} + 4H^+ + 2e^- \rightarrow SO_2(g) + 2H_2O$	+0.17
$Sn^{4+} + 2e^- \rightarrow Sn^{2+}$	+0.15
$2H^+ + 2e^- \rightarrow H_2(g)$	0.00
$Pb^{2+} + 2e^- \rightarrow Pb(s)$	−0.13
$Sn^{2+} + 2e^- \rightarrow Sn(s)$	−0.14
$Ni^{2+} + 2e^- \rightarrow Ni(s)$	−0.25
$Co^{2+} + 2e^- \rightarrow Co(s)$	−0.28
$2H^+(10^{-7}M) + 2e^- \rightarrow H_2(g)$	−0.41
$Fe^{2+} + 2e^- \rightarrow Fe(s)$	−0.44
$Cr^{3+} + 3e^- \rightarrow Cr(s)$	−0.74
$Zn^{2+} + 2e^- \rightarrow Zn(s)$	−0.76
$2H_2O + 2e^- \rightarrow 2OH^- + H_2(g)$	−0.83
$Mn^{2+} + 2e^- \rightarrow Mn(s)$	−1.18
$Al^{3+} + 3e^- \rightarrow Al(s)$	−1.66
$Mg^{2+} + 2e^- \rightarrow Mg(s)$	−2.37
$Na^+ + e^- \rightarrow Na(s)$	−2.71
$Ca^{2+} + 2e^- \rightarrow Ca(s)$	−2.87
$Sr^{2+} + 2e^- \rightarrow Sr(s)$	−2.89
$Ba^{2+} + 2e^- \rightarrow Ba(s)$	−2.90
$Cs^+ + e^- \rightarrow Cs(s)$	−2.92
$K^+ + e^- \rightarrow K(s)$	−2.92
$Rb^+ + e^- \rightarrow Rb(s)$	−2.93
$Li^+ + e^- \rightarrow Li(s)$	−3.00

24. In the reaction $Cl_2 + 2Br^-(aq) \rightarrow 2Cl^-(aq) + Br_2$, which half-reaction CORRECTLY represents oxidation? 24. ____
   A. $2Br^- \rightarrow Br_2 + 2e^-$
   B. $Cl_2 \rightarrow 2Cl^- + 2e^-$
   C. $2Br^- + 2e^- \rightarrow Br_2$
   D. $Cl_2 + 2e^- \rightarrow 2Cl^-$

25. Given the reaction: 25. ____
   $\_K_2Cr_2O_7 + \_HCl \rightarrow \_KCl + \_CrCl_3 + \_Cl_2 + \_H_2O$.
   When the reaction is completely balanced using SMALLEST whole numbers, the coefficient of Cl_2 will be
   A. 1 B. 2 C. 3 D. 4

26. Given the reaction: 26. ____
   $2Cr(s) + 3Cu^{+2}(aq) \rightarrow 2Cr^{+3}(aq) + 3Cu(s)$.
   The potential difference (E°) of this cell is _____ V.
   A. 0.40 B. 1.08 C. 1.25 D. 2.50

27. What is the CORRECT I.U.P.A.C. name for a compound with the structural formula shown at the right? 27. ____

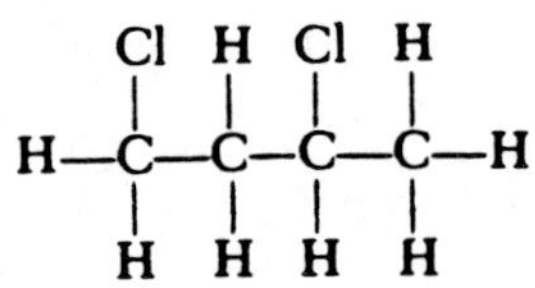

   A. 1,3-dichloropentane
   B. 2,4-dichloropentane
   C. 1,3-dichlorobutane
   D. 2,4-dichlorobutane

28. The reaction represented by the equation $nC_2H_4 \rightarrow (-C_2H_4-)_n$ is called 28. ____
   A. saponification
   B. fermentation
   C. esterification
   D. polymerization

29. Which compound is a trihydroxy alcohol? 29. ____
   A. Glycerol B. Butanol C. Ethanol D. Methanol

30. Which formula represents an acid? 30. ____

A. 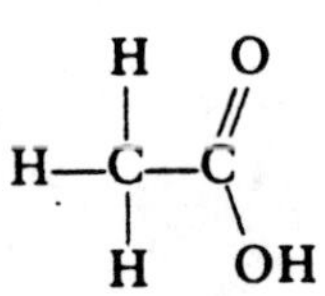

C. 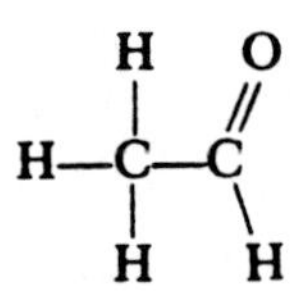

B. 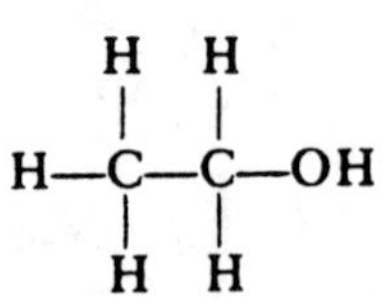

D. 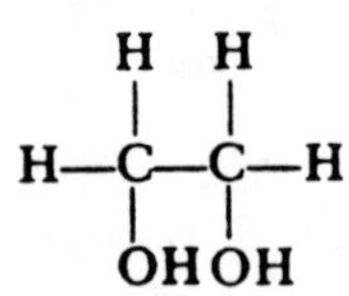

31. Organic compounds DIFFER from inorganic compounds in that organic compounds generally have _____ melting points and are _____. 31. ____
   A. high; electrolytes
   B. high; nonelectrolytes
   C. low; electrolytes
   D. low; nonelectrolytes

32. Group IA metals are obtained commercially from their fused salts by 32. ____
   A. transmutation
   B. fractional distillation
   C. electrolysis
   D. polymerization

33. When large hydrocarbon molecules are broken down into smaller hydrocarbon molecules, the reaction is known as 33.___
A. cracking
B. distilling
C. reduction
D. oxidation

34. Which equation represents the reaction that occurs during the Haber process? 34.___
A. $ZnO + C \rightarrow Zn + CO$
B. $2Al + Cr_2O_3 \rightarrow Al_2O_3 + 2Cr$
C. $S + O_2 \rightarrow SO_2$
D. $N_2 + 3H_2 \rightarrow 2NH_3$

35. Given the probable reaction for the nickel-cadmium battery: 35.___
$2Ni(OH)_3 + Cd \rightleftarrows 2Ni(OH)_2 + Cd(OH)_2$.
Which species is oxidized during the discharge of the battery?
A. Ni^{3+}
B. Ni^{2+}
C. Cd^0
D. Cd^{2+}

36. Which metal may be connected to an iron pipe to prevent the corrosion of the iron? 36.___
A. Co
B. Sn
C. Pb
D. Mg

37. The mass number of a deuterium atom is 37.___
A. one
B. two
C. three
D. four

38. In a nuclear reactor, boron and cadmium are COMMONLY used in 38.___
A. fuels
B. coolants
C. moderators
D. control rods

39. Which atom can undergo nuclear fission when its nucleus captures a neutron? 39.___
A. ${}^{1}_{1}H$
B. ${}^{2}_{1}H$
C. ${}^{235}_{92}U$
D. ${}^{238}_{92}U$

40. Consider the reaction ${}^{27}_{13}Al + {}^{4}_{2}He \rightarrow {}^{30}_{15}P + {}^{1}_{0}n$. 40.___
This reaction is BEST described as
A. beta decay
B. artificial transmutation
C. fission
D. fusion

KEY (CORRECT ANSWERS)

1. A	11. A	21. B	31. D
2. D	12. D	22. A	32. C
3. C	13. B	23. D	33. A
4. D	14. A	24. A	34. D
5. D	15. D	25. C	35. C
6. B	16. C	26. B	36. D
7. C	17. C	27. C	37. B
8. B	18. A	28. D	38. D
9. C	19. D	29. A	39. C
10. A	20. C	30. A	40. B

TEST 2

DIRECTIONS: Each question or incomplete statement is followed by several suggested answers or completions. Select the one that BEST answers the question or completes the statement. *PRINT THE LETTER OF THE CORRECT ANSWER IN THE SPACE AT THE RIGHT.*

1. Given the reaction: $A + B \rightleftarrows AB$ 1.____
The GREATEST amount of AB would be produced if the equilibrium constant of the reaction is equal to 1.0 ×
   A. 10^{-5} B. 10^{-1} C. 10^{1} D. 10^{5}

2. Based on the chart at the right, the MOST energy would be released when oxygen reacts completely with 1.0 mole of 2.____
   A. CH_4
   B. CO
   C. $C_6H_{12}O_6$
   D. CH_3OH

Heats of Reaction at 1 atm and 298K

Reaction	ΔH (kcal)
$CH_4\ (g) + 2O_2\ (g) \rightarrow CO_2\ (g) + 2H_2O\ (\ell)$	−212.8
$C_3H_8\ (g) + 5O_2\ (g) \rightarrow 3CO_2\ (g) + 4H_2O\ (\ell)$	−530.6
$CH_3OH\ (\ell) + \frac{3}{2}O_2\ (g) \rightarrow CO_2\ (g) + 2H_2O\ (\ell)$	−173.6
$C_6H_{12}O_6\ (s) + 6O_2\ (g) \rightarrow 6CO_2\ (g) + 6H_2O\ (\ell)$	−669.9
$CO\ (g) + \frac{1}{2}O_2\ (g) \rightarrow CO_2\ (g)$	−67.7
$NaOH\ (s) \xrightarrow{H_2O} Na^+\ (aq) + OH^-\ (aq)$	−10.6
$NH_4Cl\ (s) \xrightarrow{H_2O} NH_4^+\ (aq) + Cl^-\ (aq)$	+3.5
$H^+\ (aq) + OH^-\ (aq) \rightarrow H_2O\ (\ell)$	−13.8

3. At which pressure would carbon dioxide gas be MOST soluble in 100 grams of water at a temperature of 25°C? 3.____
_____ atm.
   A. 1 B. 2 C. 3 D. 4

4. According to the table at the right, which species can act as both a Brönsted acid and a Brönsted base? 4.____
   A. NO_2^-
   B. NO_3^-
   C. HF
   D. NH_3

RELATIVE STRENGTHS OF ACIDS IN AQUEOUS SOLUTION at 1 atm AND 298 K

Conjugate Pairs ACID → BASE	K_a
$HI \rightarrow H^+ + I^-$	very large
$HBr \rightarrow H^+ + Br^-$	very large
$HCl \rightarrow H^+ + Cl^-$	very large
$HNO_3 \rightarrow H^+ + NO_3^-$	very large
$H_2SO_4 \rightarrow H^+ + HSO_4^-$	large
$H_2O + SO_2 \rightarrow H^+ + HSO_3^-$	1.7×10^{-2}
$HSO_4^- \rightarrow H^+ + SO_4^{2-}$	1.3×10^{-2}
$H_3PO_4 \rightarrow H^+ + H_2PO_4^-$	7.1×10^{-3}
$Fe(H_2O)_6^{3+} \rightarrow H^+ + Fe(H_2O)_5(OH)^{2+}$	6.0×10^{-3}
$HF \rightarrow H^+ + F^-$	6.7×10^{-4}
$HNO_2 \rightarrow H^+ + NO_2^-$	5.1×10^{-4}
$Cr(H_2O)_6^{3+} \rightarrow H^+ + Cr(H_2O)_5(OH)^{2+}$	1.0×10^{-4}
$CH_3COOH \rightarrow H^+ + CH_3COO^-$	1.8×10^{-5}
$Al(H_2O)_6^{3+} \rightarrow H^+ + Al(H_2O)_5(OH)^{2+}$	1.0×10^{-5}
$H_2O + CO_2 \rightarrow H^+ + HCO_3^-$	4.4×10^{-7}
$H_2S \rightarrow H^+ + HS^-$	1.0×10^{-7}
$H_2PO_4^- \rightarrow H^+ + HPO_4^{2-}$	6.3×10^{-8}
$HSO_3^- \rightarrow H^+ + SO_3^{2-}$	6.2×10^{-8}
$NH_4^+ \rightarrow H^+ + NH_3$	5.7×10^{-10}
$HCO_3^- \rightarrow H^+ + CO_3^{2-}$	4.7×10^{-11}
$HPO_4^{2-} \rightarrow H^+ + PO_4^{3-}$	4.4×10^{-13}
$HS^- \rightarrow H^+ + S^{2-}$	1.3×10^{-13}
$OH^- \rightarrow H^+ + O^{2-}$	$< 10^{-36}$
$NH_3 \rightarrow H^+ + NH_2^-$	very small

5. The ionization constant, K_a, for acetic acid at 1 atmosphere and 298 K is 5.___
   A. 2.5×10^{-11}
   B. 2.1×10^{-8}
   C. 1.8×10^{-5}
   D. 1.3×10^{-2}

6. The H_3O^+ ion concentration of a solution is 1×10^{-5} mole per liter. 6.___
   This solution is _____ and has a pH of _____.
   A. acidic; 5
   B. acidic; 9
   C. basic; 5
   D. basic; 9

7. According to the Arrhenius theory, a substance that yields hydrogen ions as the only positive ion in an aqueous solution is a(n) 7.___
   A. salt
   B. base
   C. acid
   D. nonelectrolyte

8. How many moles of HCl can be neutralized by 0.1 liter of 0.5 M NaOH? 8.___
   A. 0.1
   B. 0.05
   C. 0.5
   D. 0.4

9. A 0.1 M solution of HCl contains 9.___
   A. fewer H_3O^+ ions than OH^- ions
   B. more H_3O^+ ions than OH^- ions
   C. an equal number of H_3O^+ and OH^- ions
   D. neither H_3O^+ ions nor OH^- ions

10. In the reaction $Cl_2 + H_2O \rightarrow HClO + HCl$, the oxidation number of chlorine 10.___
   A. decreases *only*
   B. increases *only*
   C. neither decreases nor increases
   D. both decreases and increases

11. According to the chart at the right, which can act as both a reducing agent and an oxidizing agent? 11.___

A. Au^{3+}
B. Al^{3+}
C. Fe^{2+}
D. Zn^{2+}

STANDARD ELECTRODE POTENTIALS	
Ionic Concentrations 1 M Water at 298 K, 1 atm	
Half-Reaction	E° (volts)
$F_2(g) + 2e^- \rightarrow 2F^-$	+2.87
$MnO_4^- + 8H^+ + 5e^- \rightarrow Mn^{2+} + 4H_2O$	+1.52
$Au^{3+} + 3e^- \rightarrow Au(s)$	+1.50
$Cl_2(g) + 2e^- \rightarrow 2Cl^-$	+1.36
$Cr_2O_7^{2-} + 14H^+ + 6e^- \rightarrow 2Cr^{3+} + 7H_2O$	+1.33
$MnO_2(s) + 4H^+ + 2e^- \rightarrow Mn^{2+} + 2H_2O$	+1.28
$\frac{1}{2}O_2(g) + 2H^+ + 2e^- \rightarrow H_2O$	+1.23
$Br_2(\ell) + 2e^- \rightarrow 2Br^-$	+1.06
$NO_3^- + 4H^+ + 3e^- \rightarrow NO(g) + 2H_2O$	+0.96
$\frac{1}{2}O_2(g) + 2H^+(10^{-7}M) + 2e^- \rightarrow H_2O$	+0.82
$Ag^+ + e^- \rightarrow Ag(s)$	+0.80
$\frac{1}{2}Hg_2^{2+} + e^- \rightarrow Hg(\ell)$	+0.79
$Hg^{2+} + 2e^- \rightarrow Hg(\ell)$	+0.78
$NO_3^- + 2H^+ + e^- \rightarrow NO_2(g) + H_2O$	+0.78
$Fe^{3+} + e^- \rightarrow Fe^{2+}$	+0.77
$I_2(s) + 2e^- \rightarrow 2I^-$	+0.53
$Cu^+ + e^- \rightarrow Cu(s)$	+0.52
$Cu^{2+} + 2e^- \rightarrow Cu(s)$	+0.34
$SO_4^{2-} + 4H^+ + 2e^- \rightarrow SO_2(g) + 2H_2O$	+0.17
$Sn^{4+} + 2e^- \rightarrow Sn^{2+}$	+0.15
$2H^+ + 2e^- \rightarrow H_2(g)$	0.00
$Pb^{2+} + 2e^- \rightarrow Pb(s)$	−0.13
$Sn^{2+} + 2e^- \rightarrow Sn(s)$	−0.14
$Ni^{2+} + 2e^- \rightarrow Ni(s)$	−0.25
$Co^{2+} + 2e^- \rightarrow Co(s)$	−0.28
$2H^+(10^{-7}M) + 2e^- \rightarrow H_2(g)$	−0.41
$Fe^{2+} + 2e^- \rightarrow Fe(s)$	−0.44
$Cr^{3+} + 3e^- \rightarrow Cr(s)$	−0.74
$Zn^{2+} + 2e^- \rightarrow Zn(s)$	−0.76
$2H_2O + 2e^- \rightarrow 2OH^- + H_2(g)$	−0.83
$Mn^{2+} + 2e^- \rightarrow Mn(s)$	−1.18
$Al^{3+} + 3e^- \rightarrow Al(s)$	−1.66
$Mg^{2+} + 2e^- \rightarrow Mg(s)$	−2.37
$Na^+ + e^- \rightarrow Na(s)$	−2.71
$Ca^{2+} + 2e^- \rightarrow Ca(s)$	−2.87
$Sr^{2+} + 2e^- \rightarrow Sr(s)$	−2.89
$Ba^{2+} + 2e^- \rightarrow Ba(s)$	−2.90
$Cs^+ + e^- \rightarrow Cs(s)$	−2.92
$K^+ + e^- \rightarrow K(s)$	−2.92
$Rb^+ + e^- \rightarrow Rb(s)$	−2.93
$Li^+ + e^- \rightarrow Li(s)$	−3.00

12. Given the reaction: $\_Fe^{3+} + \_Sn^{2+} \rightarrow \_Fe^{2+} + \_Sn^{4+}$. When the reaction is completely balanced using SMALLEST whole numbers, the coefficient of Fe^{+} will be 12.___

A. 1 B. 2 C. 3 D. 4

13. In the reaction $Zn + Cu^{2+} \rightarrow Zn^{2+} + Cu$, the Cu^{2+} ions 13.___

A. gain electrons B. lose electrons
C. gain protons D. lose protons

14. In the reaction $2K + Cl_2 \rightarrow 2KCl$, the species oxidized is 14.___

A. Cl_2 B. Cl^- C. K D. K^+

15. Indicate which of the following compounds will undergo a substitution reaction with chlorine. 15.___

A. CH_4 B. C_2H_4 C. C_3H_6 D. C_4H_8

16. Which formula represents a hydrocarbon with a double covalent bond? 16.___

A. CH_3Cl B. C_2H_3Cl C. C_2H_2 D. C_2H_4

17. Which compound is an isomer of 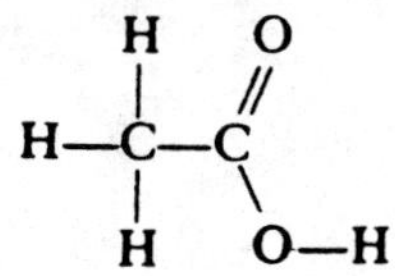 ? 17.____

A. 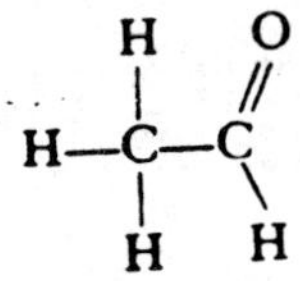

B. 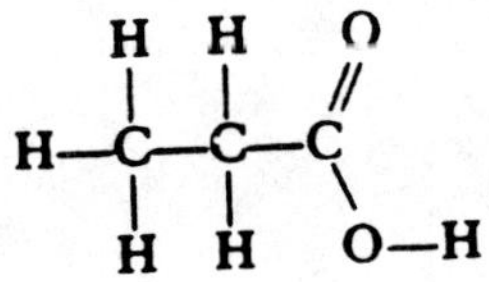

C. 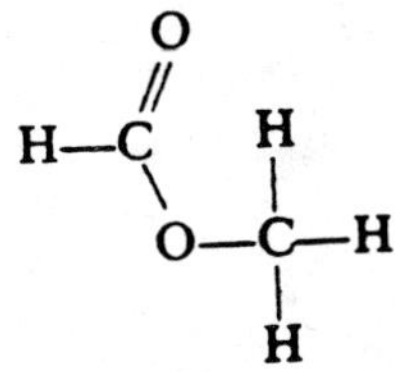

D. 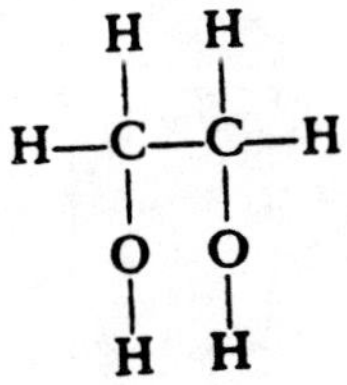

18. The TOTAL number of covalent bonds in a molecule of methane is 18.____

A. one B. two C. three D. four

19. As an S^{2-} ion is oxidized to an S^0 atom, the number of protons in its nucleus 19.____

A. decreases
B. increases
C. remains the same
D. one cannot tell from the information given

20. As 1 gram of $H_2O(g)$ changes to 1 gram of $H_2O(\ell)$, the entropy of the system 20.____

A. decreases
B. increases
C. remains the same
D. one cannot tell from the information given

21. The volume of a sample of hydrogen gas at STP is 1.00 liter. As the temperature decreases, pressure remaining constant, the volume of the sample 21.____

A. decreases
B. increases
C. remains the same
D. one cannot tell from the information given

Questions 22-23.

DIRECTIONS: Questions 22 and 23 are to be answered on the basis of the graph below. The graph shows the relationship between temperature and time as heat is added to one mole of a substance at a rate of 100 calories per minute. The substance is in the solid phase at Time = 0 minutes.

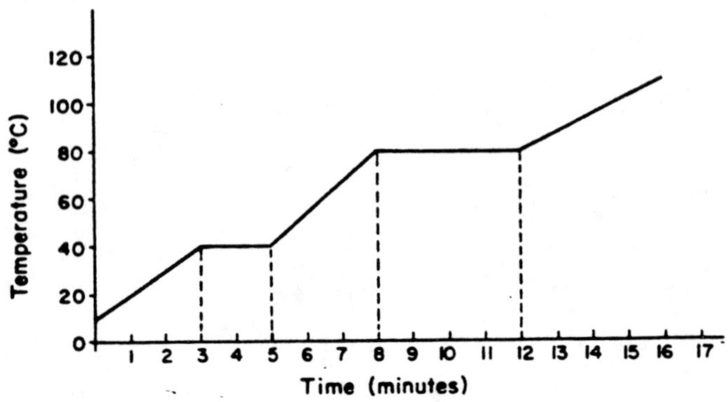

22. The temperature at which the substance begins to boil is ______ °C. 22.___
    A. 10 B. 40 C. 80 D. 110

23. From the time that the solid begins to melt, the MINIMUM number of calories required to completely melt the one mole sample is 23.___
    A. 100 B. 200 C. 400 D. 600

24. The pressure on 200. milliliters of a gas at constant temperature is changed from 380. torr to 760. torr. The NEW volume of the gas is ______ ml. 24.___
    A. 100. B. 200. C. 400. D. 800.

25. The temperature of a substance changes from -173°C to 0°C. This change represents ______ Kelvin degrees. 25.___
    A. 100. B. 173 C. 273 D. 446

26. How many kilocalories of heat are needed to raise the temperature of 500. grams of water from 10.0°C to 30.0°C? ______ kcal. 26.___
    A. 10.0 B. 25.0 C. 50.0 D. 40.0

27. Isotopes of the same element must have the same 27.___
    A. atomic number
    B. mass number
    C. number of nucleons
    D. number of neutrons

28. Which pair has identical electron configurations? 28.___
    A. S^{2-} and Cl^-
    B. S^0 and Ar^0
    C. K^0 and Na^+
    D. Cl^- and K^0

29. The energy level with a principal quantum number (n) of 2 contains a total of _____ sublevels. 29.___
A. eight B. two C. three D. four

30. What is the TOTAL mass of a 10. gram sample of ^{42}K that will remain unchanged after 12.4 hours? 30.___
_____ grams.
A. 2.5 B. 5.0 C. 7.5 D. 10.

31. Which particle has a mass that is APPROXIMATELY 1/1,836 of the mass of a proton? 31.___
A(n)
A. gamma ray B. neutron
C. alpha particle D. electron

32. The P-Cl bond in a molecule of PCl_3 is 32.___
A. nonpolar covalent B. polar covalent
C. coordinate covalent D. electrovalent

33. The formula H_2 represents one 33.___
A. gram B. liter C. atom D. molecule

34. The predominate type of attraction between molecules of HF in the liquid state is _____ bonding. 34.___
A. hydrogen B. electrovalent
C. ionic D. covalent

35. Which substance contains positive ions immersed in a sea of mobile electrons? 35.___
A. $O_2(s)$ B. $SiO_2(s)$ C. Cu(s) D. CuO(s)

36. The radioactive isotope I-131 is used in 36.___
A. geologic dating
B. medical diagnosis
C. controlling fission reactions
D. controlling speeds of neutrons

37. Which measurement contains a total of three significant figures? 37.___
_____ grams.
A. 0.01 B. 0.010 C. 0.0100 D. 0.01000

38. The diagram at the right represents a portion of a triple beam balance. If the beams are in balance, with the riders in the position shown, what is the TOTAL mass in grams of the object being massed? 38.___

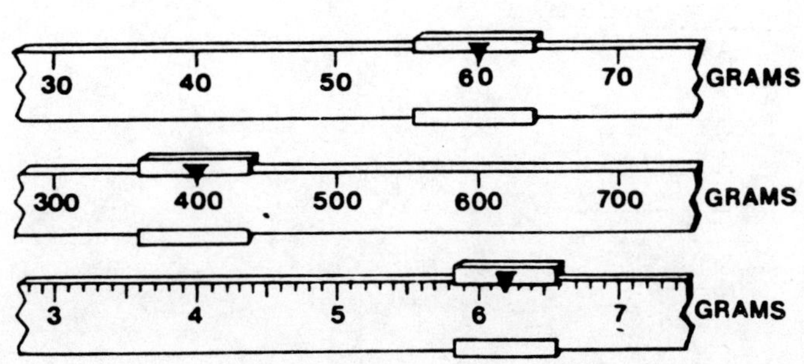

A. 460.62
B. 466.20
C. 466.62
D. 460.20

39. According to an accepted chemistry reference, the heat of vaporization of water is 540. calories per gram. A student determined in the laboratory that the heat of vaporization of water was 620. calories per gram. The student's results had a percent error of 39.___
A. 12.9 B. 14.8 C. 80.0 D. 87.1

40. The data below were obtained by a student in order to determine the percent of water in a hydrate: 40.___

Mass of the Hydrate	5.0 g
Mass of the Anhydrous Compound	3.2 g

The hydrate contains _____ percent water.
A. 64 B. 56 C. 36 D. 22

KEY (CORRECT ANSWERS)

1. D	11. C	21. A	31. D
2. C	12. B	22. C	32. B
3. D	13. A	23. B	33. D
4. D	14. C	24. A	34. A
5. C	15. A	25. B	35. C
6. A	16. D	26. A	36. B
7. C	17. C	27. A	37. C
8. B	18. D	28. A	38. B
9. B	19. C	29. B	39. B
10. D	20. A	30. B	40. C

TEST 3

DIRECTIONS: Each question or incomplete statement is followed by several suggested answers or completions. Select the one that BEST answers the question or completes the statement. *PRINT THE LETTER OF THE CORRECT ANSWER IN THE SPACE AT THE RIGHT.*

1. Element A and element B chemically combine to form substance C. Substance C MUST be a(n) 1.___
   A. solution B. compound C. element D. mixture

2. Which gas has properties that are MOST similar to those of an ideal gas? 2.___
   A. O_2 B. H_2 C. NH_3 D. HCl

3. When the vapor pressure of a liquid equals the pressure on its surface, the liquid will 3.___
   A. sublime B. solidify C. boil D. melt

4. The energy required to change a unit mass of a solid to a liquid at constant temperature is called its heat of 4.___
   A. formation B. vaporization
   C. combustion D. fusion

5. A 1 liter flask of $CO_2(g)$ and a 1 liter flask of $H_2(g)$ are both at STP. The ratio of the number of molecules of $CO_2(g)$ to the number of molecules of $H_2(g)$ in the flasks is 5.___
   A. 1:1 B. 2:3 C. 3:2 D. 1:3

6. The number of protons in the nucleus of $^{32}_{15}P$ is 6.___
   A. 15 B. 17 C. 32 D. 47

7. Atomic mass is measured in atomic mass units (amu) that are based on an atom of ______ equal to ______ amu. 7.___
   A. ^{16}O; 16.000 B. ^{32}S; 32.000
   C. ^{12}C; 12.000 D. ^{14}N; 14.000

8. According to the chart at the right, which change requires the LEAST amount of energy? 8.___
   A. $Li^0(g) \rightarrow Li^+(g) + e^-$
   B. $Rb^0(g) \rightarrow Rb^+(g) + e^-$
   C. $K^0(g) \rightarrow K^+(g) + e^-$
   D. $Na^0(g) \rightarrow Na^+(g) + e^-$

Ionization Energies and Electronegativities

Upper number: First Ionization Energy (kcal/mole of atoms); lower number: Electronegativity

IA	IIA	IIIA	IVA	VA	VIA	VIIA	0
H 313 / [illegible]							He 567
Li 124 / 1.0	Be 215 / 1.5	B 191 / 2.0	C 260 / 2.5	N 336 / 3.0	O 314 / 3.5	F 402 / 4.0	Ne 497
Na 119 / 0.9	Mg 176 / 1.2	Al 138 / 1.5	Si 188 / 1.8	P 254 / 2.1	S 239 / 2.5	Cl 300 / 3.0	Ar 363
K 100 / 0.8	Ca 141 / 1.0	Ga 138 / 1.6	Ge 187 / 1.8	As 231 / 2.0	Se 225 / 2.4	Br 273 / 2.8	Kr 323
Rb 96 / 0.8	Sr 131 / 1.0	In 133 / 1.7	Sn 169 / 1.8	Sb 199 / 1.9	Te 208 / 2.1	I 241 / 2.5	Xe 280
Cs 90 / 0.7	Ba 120 / 0.9	Tl 141 / 1.8	Pb 171 / 1.8	Bi 185 / 1.9	Po / 2.0	At / 2.2	Rn 248
Fr / 0.7	Ra						

9. An atom has the electron configuration $1s^2 2s^2 2p^6 3s^2 3p^5$. The electron dot symbol for this element is 9.____

A. X: B. X: C. ·X: D. :X:

10. Which sublevel has a total of five orbitals? 10.____

A. s B. f C. p D. d

11. The MAXIMUM number of electrons that may be found in the third principal energy level is 11.____

A. 8 B. 2 C. 18 D. 32

12. According to the chart at the right, a product of the radioactive decay of $^{226}_{88}Ra$ is 12.____

A. $^{4}_{2}He$

B. $^{226}_{89}U$

C. $^{0}_{-1}e$

D. $^{230}_{90}U$

SELECTED RADIOISOTOPES		
Nuclide	*Half-Life*	*Particle Emission*
^{14}C	5730 y	β^-
^{60}Co	5.3 y	β^-
^{137}Cs	30.23 y	β^-
^{220}Fr	27.5 s	α
^{3}H	12.26 y	β^-
^{131}I	8.07 d	β^-
^{40}K	1.28×10^9 y	β^+
^{42}K	12.4 h	β^-
^{32}P	14.3 d	β^-
^{226}Ra	1600 y	α
^{90}Sr	28.1 y	β^-
^{235}U	7.1×10^8 y	α
^{238}U	4.51×10^9 y	α

y = years; d = days;
h = hours; s = seconds

13. In the balanced equation $2Na + 2H_2O \rightarrow H_2 + 2X$, the compound represented by X is 13.____

A. Na_2O B. Na_2O_2 C. NaOH D. NaH

14. Indicate which of the following elements has the LOWEST normal boiling point 14.____

A. Ne B. He C. Ar D. Kr

15. The CORRECT formula for sodium oxide is 15.____

A. SO_2 B. S_2O C. NaO_2 D. Na_2O

16. The attraction which exists between carbon dioxide molecules in solid carbon dioxide is due to 16.____

A. van der Waals forces B. molecule-ion forces
C. ionic bonds D. hydrogen bonds

17. Given the reaction: 17.____

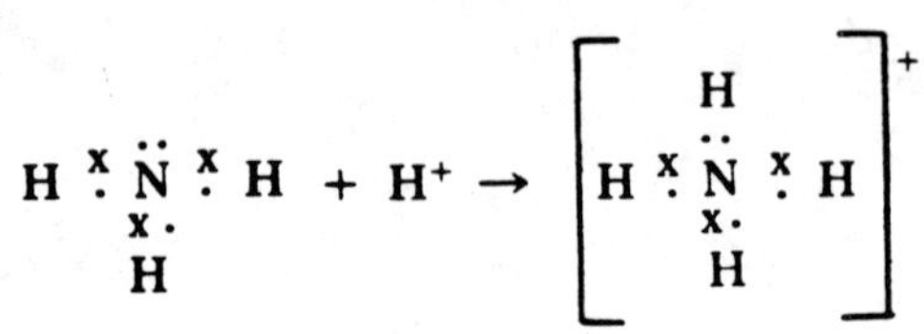

The bond formed between the NH_3 and the H^+ is

A. electrovalent
B. ionic
C. metallic
D. coordinate covalent

18. Which of the following compounds has the LEAST ionic character? 18.___
A. KI B. NO C. HCl D. MgS

19. What type of bond is present in diamonds? 19.___
A. Covalent B. Ionic
C. Electrovalent D. Metallic

20. All the elements in Period 3 have the same number of 20.___
A. occupied sublevels B. principal energy levels
C. electrons D. protons

21. More than two-thirds of the elements are classified as 21.___
A. nonmetals B. metals
C. metalloids D. noble gases

22. Which element in Group VA has the GREATEST metallic character? 22.___
A. N B. P C. Sb D. Bi

23. The element in Period 3 that has the LOWEST first ionization energy is a(n) 23.___
A. alkali metal B. alkaline earth metal
B. halogen D. noble gas

24. When fluorine reacts with a Group IA metal, it becomes an ion with a charge of 24.___
A. 1^- B. 2^- C. 1^+ D. 2^+

25. The alkaline earth metals are in Group 25.___
A. IA B. IB C. IIA D. IIB

26. Which of the following metals has the LOWEST melting point? 26.___
A. Copper B. Mercury C. Silver D. Iron

27. A compound contains 0.5 mole of sodium, 0.5 mole of nitrogen, and 1.0 mole of hydrogen. 27.___
The empirical formula of the compound is
A. NaNH B. Na_2NH C. $NaNH_2$ D. $Na(NH)_2$

28. The percent by mass of oxygen in CO is APPROXIMATELY ______ percent. 28.___
A. 73 B. 57 C. 43 D. 17

29. The mass in grams of 1.00 mole of $CaSO_4 \cdot 2H_2O$ is ____ grams. 29.___
A. 172 B. 154 C. 136 D. 118

30. The total number of sodium atoms in 46.0 grams of sodium is ______ $\times 10^{23}$. 30.___
A. 3.01 B. 6.02 C. 12.0 D. 24.0

31. Given the reactions: 31.___

$4NH_3(g) + 5O_2(g) \rightarrow 4NO(g) + 6H_2O(g)$.

At STP, what is the TOTAL volume of $O_2(g)$ required to produce 80.0 liters of NO(g)?

_____ ℓ.

A. 5.00 B. 20.0 C. 80.0 D. 100.

32. What is the TOTAL number of grams of solute in 500. milliliters of 1 M CH_3COOH (formula mass = 60.)? 32.___

_____ grams.

A. 30. B. 60. C. 90. D. 120

33. At 10°C, 23 grams of a substance saturates 100 grams of water. 33.___

The substance could be

A. $NaNO_3$ B. KNO_3 C. NH_4Cl D. KCl

34. The addition of a catalyst to a reaction will cause a change in the 34.___

A. potential energy of the reactants
B. potential energy of the products
C. heat of reaction
D. activation energy

35. Given the reaction at equilibrium: 35.___

$A(g) + B(g) \rightleftarrows C(g) + D(g)$.

At constant temperature and pressure, an increase in the concentration of A(g) will cause a(n) _____ in the concentration of _____.

A. increase; B(g) B. decrease; B(g)
C. decrease; C(g) D. decrease; D(g)

KEY (CORRECT ANSWERS)

1. B	11. C	21. A	31. D
2. B	12. A	22. D	32. A
3. C	13. C	23. A	33. C
4. D	14. B	24. A	34. D
5. A	15. D	25. C	35. B
6. A	16. A	26. B	
7. C	17. D	27. C	
8. B	18. B	28. B	
9. D	19. A	29. A	
10. D	20. B	30. C	

CHEMICAL NOTES AND RESOURCES

CONTENTS

CHEMICAL NOTES AND RESOURCES

GLOSSARY OF CHEMICAL TERMS

A

ABSOLUTE TEMPERATURE
Temperature on the Absolute scale whose zero is -273°Centigrade.

ABSORPTION
A soaking up throughout the mass.

ACCELERATOR
Chemical additive which hastens or increases the speed of a chemical reaction. Used,for example,to improve the vulcanization of natural and synthetic rubber and latex compounds.

ACID
A water soluble chemical compound containing hydrogen replaceable by metals or basic radicals. An acid reacts with an alkali to form a salt and water. Example of acid: sulfuric acid,commonly used in storage batteries and many other applications.

ACID ANHYDRIDE
An oxide of a nonmetal capable of uniting with water to form an acid.

ACID-FORMERS
A property usually characteristic of those elements called nonmetals.

ACID SALT
A salt containing replaceable hydrogen.

ACID SULFITE
A salt formed by the union of a metal with the HSO_3 ion.

ACRILAN
One of the synthetic fibers.

ADSORPTION
A condensation on the surface of a material.

ADSORPTION, ACTIVATED
That form of adsorption in which sufficient heat energy must be supplied before the film forms. This process is probably chemical in nature.

AERATED
Given an opportunity to dissolve or combine with air or some other gas.

AEROSOLS
Fog-like sprays.

AIR-SLAKED LIME
A mixture of calcium hydroxide and calcium carbonate formed by exploding calcium hydroxide to the air.

ALCOHOL
A compound containing an organic radical and one or more OH groups.

ALDEHYDE
An organic compound containing the CHO group.

ALIPHATIC
(Derived from Greek word for fat.) Pertaining to an open chain carbon compound. Usually applied to petroleum products derived from a paraffin base and having a straight or branched chain,saturated or unsaturated, molecular structure as distinguished from aromatic hydrocarbons which are built up from one or more benzene rings. Gasoline is a typical aliphatic hydrocarbon.

ALKALI
A compound that has the power to neytralize an acid and form a salt. Example: sodium hydroxide,referred to as caustic soda or lye. Used in soap manufacture and many other applications.

ALKYLATION
A process for rearranging straight chain hydrocarbons.

ALLOTROPISM
The ability of some elements to exist in more than one form.
ALLOY
A material composed of two or more metals.
ALPHA PARTICLES
Positively charged helium nuclei.
ALUM
The double sulfate of a monovalent and trivalent metal, containing a definite amount of water of hydration ($KAl(SO_4)_2 \cdot 12H_2O$).
ALUMINOTHERMY
A thermite or similar reaction.
AMALGAM
An alloy of mercury with one or more other metals.
AMALGAMATION PROCESS
A process of extracting gold from ore by amalgamating the gold with mercury.
AMINO ACIDS
The "building blocks" from which the giant protein molecules are constructed.
AMMONIA (NH_3)
Nitrogen and hydrogen compound, a colorless gas liquefied by compression. Dissolves in water to form aqueous ammonia. Synthetic ammonia is main source of nitrogen for fertilizer and chemical production.
AMMONIUM ION, NH_4^+
This is a cation produced by the ionization of an ammonium salt.
AMMONIUM RADICAL, NH_4
An ammonium radical is a group of atoms which plays the role of a metal in certain salts (e.g., NH_4Cl).
AMORPHOUS
This is a substance without crystalline structure. The atoms or molecules are not arranged in a definite pattern.
AMPERE
An Ampere is one coulomb of electricity per second. That current which deposits .001118g. silver per second.
AMPHOTERIC
Referring to a compound which may ionize as a base in the presence of a strong acid, and as an acid in the presence of a strong base.
ANGSTROM UNIT (A)
$= 10^{-8}$ cm.
ANHYDRIDE
A compound, usually an oxide of a metal or a nonmetal, capable of uniting with water to form a base or an acid.
ANHYDROUS
Material from which the water has been removed.
ANILINE ($C_6H_5NH_2$)
One of the most important of organics derived from coal. Building block for many dyes and drugs.
ANION
An anion is a negatively charged ion. It is attracted to the anode (+electrode) during electrolysis.
ANODE
The positive terminal of an electric cell.
ANTIBIOTIC
A substance either synethesized chemically or produced by a microorganism, usually a mold or fungus, which kills other organisms, or retards, or completely represses their growth, normally without harm to higher orders of life. Antibiotics retain highly germicidial properties even in dilute concentrations.

ANTICHLOR
A chemical which acts against chlorine, such as a solution of sodium thiosulfate.

ANTIOXIDANT
A compound added to rubber and other substances to prevent deterioration by oxidation.

AQUA AMMONIA
A solution of ammonia gas in water.

AQUA REGIA
A mixture of concentrated nitric and hydrochloric acids capable of dissolving gold.

ARC-TYPE FURNACE
One that has an electric arc jumping the gap between carbon electrodes.

AROMATIC
Applied to group of hydrocarbons derived from or characterized by presence of the benzene nucleus (molecular ring structure). Sometimes called "cyclic solvents" or "cyclic hydrocarbons."

ASSOCIATION
A joining together of small molecules to form larger molecules.

ATOM
A chemical unit, the smallest part of an element which remains unchanged during any chemical reaction yet may undergo physical changes to other atoms as in atomic fission. Believed to be made up of a complex system whose electrically charged components are in rapid orbital motion.

ATOMIC NUMBER
A number identifying an element, equal to the number of protons in its atoms.

ATOMIC WEIGHT
The average relative weight of the atoms of an element compared with those of oxygen taken as a standard and given a value of 16.

B

BACITRACIN
One of the newer antibiotics.

BAKING SODA
The compound sodium hydrogen carbonate used in baking powders and for other purposes

BAROMETER
An apparatus for measuring atmospheric pressure.

BASE
A compound containing the hydroxyl group which, when dissolved in water, forms no negative ions but hydroxyl ions.

BASE-FORMERS
A property usually characteristic of metals.

BASIC ANHYDRIDE
An oxide of a metal capable of reacting with water to form a base.

BASIC SALT
A salt containing replaceable oxygen or hydroxyl groups.

BENZENE (C_6H_6)
Major organic intermediate derived from coal or petroleum. Ring-shaped (cyclic) molecular structure makes it broadly useful as chemical building block.

BESSEMER CONVERTER
An egg-shaped converter which changes pig iron into steel by burning out impurities with an air blast.

BETA PARTICLES
Negatively charged particles, actually electrons, emitted by some radioactive materials.
BETATRON
One of the types of "atom-smashing" machines for bombarding atomic nuclei.
BINARY
Binary compounds are made up of only two elements.
BISQUE
A porous porcelain product that has been fired only once.
BLOCK TIN
Solid tin, as distinguished from tin-plate.
BOILING POINT
The temperature at which the vapor pressure of a liquid reaches atmospheric pressure.
BOTTLED GAS
Propane, butane, or a mixture of both, stored under pressure in steel cylinders and used as a fuel.
BREEDER REACTOR
A reactor which uses some fissionable material to produce energy and a greater quantity of fissionable material.
BRITISH THERMAL UNIT (B.T.U.)
The quantity of heat necessary to raise the temperature of one pound of water one degree Fahrenheit.
BROWNIAN MOVEMENT
The zigzag movement of colloidal particles through the medium in which they are suspended.
BUFFER
A suitable mixture of salt and acid (or salt and base) that regulates or stabilizes the pH of a solution.

C

CALCINE
A partially refined copper ore.
CALORIE
A unit for measuring heat, equal to the amount of heat necessary to raise one gram of water one degree Centigrade.
CALORIMETER
A vessel used in measuring the heat evolved in chemical or physical changes.
CARAT
A unit of weight used for gems, equal to 200 milligrams.
CARBOGEN
A mixture of oxygen with 5% to 10% of carbon dioxide.
CARBOHYDRATES
Compounds containing carbon, hydrogen, and oxygen, usually with the hydrogen and oxygen present in the ratio of two to one.
CARBONATED BEVERAGES
Beverages which contain dissolved carbon dioxide.
CARBOXYL
The COOH group in organic compounds as found in organic acids.
CARBURETOR
The second or middle chamber of a water gas apparatus in which the gas is often enriched by spraying in oil or by adding propane.

CATALYST

A substance which through acceleration or retardation changes the spread of a chemical reaction and effects a definite change in composition and/or properties of the end product. In paint manufacture, catalysts generally become part of the final product. In most uses, however, they do not.

CATHODE

The negative electrode or terminal.

CATION

(1) A positively charged ion. (2) The ion attracted to the cathode in electrolysis.

CAUSTIC

A substance that attacks skin, hair, or such materials by chemical action.

CELLULOSE ($C_6H_{10}O_{5n}$)

A carbohydrate which makes up the structural material of vegetable tissues and fibers. Purest forms: chemical cotton and chemical pulp. Basis of rayon, acetate and cellophane.

CEMENT

A mixture made from limestone and clay which, after mixing with water, sets to a hard mass.

CEMENTATION

A process formerly used for making steel by heating wrought iron in red-hot charcoal for a long period of time.

CENTIGRADE TEMPERATURE

Temperature on the Centigrade scale which has 0° for the freezing point of water and 100° for the boiling point at a pressure of 760 mm.

CHAIN REACTION

A reaction in which the material or energy which starts the reaction is also a product.

CHAMBER PROCESS

A process for making sulfuric acid in large lead chambers, using oxides of nitrogen to promote the necessary reaction.

CHECKERWORK

Loosely-stacked firebricks in a chamber providing a circuitous passage for fuel gas or air.

CHEMICAL CHANGE

A change which produces a new substance with new properties.

CHEMICAL EQUATION

A qualitative and quantitative expression of a chemical change.

CHEMICAL EQUILIBRIUM

A reaction in which the products unite to form the original reactants at the same speed at which the reactants are forming the product.

CHEMICAL SYMBOL

Either one or two letters used as an abbreviation for an element.

CHEMOTHERAPY

Use of chemicals of particular molecular structure in the treatment of specific disorders on the assumption that known structures exhibit an affinity for certain parts of cells of affected tissues and thereby eliminate the causative factors.

CHEMURGY

That branch of applied chemistry devoted to industrial utilization of organic raw materials, especially farm products, as in the use of pine tree cellulose for rayon and paper, and soy bean oil for paints and varnishes.

CHLORINATION

Adding chlorine to a material.

CHLOROMYCETIN
One of the antibiotics.
CHLOROPHYLL
Green coloring matter in leaves which acts as a catalyst for photosynthesis.
CHLORTETRACYCLINE
One of the antibiotics, formerly called aureomycin.
CHROMOPHORS
Certain groups whose presence results in compounds having a color.
COAL GAS
A fuel gas obtained by the destructive distillation of soft coal.
COKE OVEN CHEMICALS
Those organic compounds derived from bituminous coal in the production of metallurgical coke. This major chemical raw materials source provides a base for thousands of chemicals.
COLLOID
A particle in an extremely fine state of subdivision.
COMBINATION REACTIONS
Those in which one element reacts with another to form a compound.
COMBINING WEIGHT
The number of grams of an element that will combine with or replace 8 grams of oxygen or its equivalent.
COMBUSTION
Oxidation accompanied by noticeable light and heat.
COMMON-ION EFFECT
The addition of a substance containing an ion common to that already present,causing the reaction to be driven in a definite direction.
COMPONENT
One of a minimum number of substances necessary to give the composition of a system.
COMPOST
A product of the decay of plant material.
COMPOUND
A substance composed of two or more elements joined according to the laws of chemical combination. Each compound has its own characteristic properties different from those of its elements.
CONCENTRATED
Containing much of a material, the opposite of dilute.
CONDENSATION
(1) Changing a material such as vapor to a liquid;
(2) Increasing the size of very small particles up to colloidal size;
(3) A reaction between raw materials in the making of a plastic that results in the formation of water as one of the products.
CONTACT PROCESS
A process for making sulfuric acid in which the sulfur dioxide and oxygen come in contact with a catalyst.
COORDINATE VALENCE
A kind of chemical bonding somewhat similar to co-valence,but in which there is only one donor atom.
CORTISONE
A compound of animal origin, used particularly for athritis.
COSMIC RAYS
These are rays which come to the earth from somewhere in space, perhaps beyond the solar system.
COTTRELL PRECIPITATOR
A device for precipitating colloidal dust with electricity of high voltage.

COUNTERCURRENTS
Two currents proceeding in opposite directions through an apparatus.
COVALENCE
A kind of chemical bonding in which two atoms share an electron with each other.
C.P.
Abbreviation for "chemically pure."
CRACKING
Breaking large or complex molecules into simpler, smaller molecules.
CRITICAL PRESSURE
The pressure of a system at its critical temperature.
CRITICAL SIZE
The smallest amount of fissionable material which can sustain a chain reaction.
CRITICAL TEMPERATURE
The highest temperature at which a liquid and its vapor can co-exist as separate phases.
CRYSTALLOID
A term applied to materials that crystallize easily and pass through a semi-permeable membrane without difficulty.
CRYSTALS
Solids separated from solutions having a definite shape or structure.
CULLET
Broken glass that is re-melted with raw materials in a new batch of glass.
CUPELLATION
A process of separating gold and silver from a base metal such as lead. An oxidizing flame converts the lead to the oxide(PbO), which is removed by a stream of air or is absorbed in the porous bottom of the reverberatory furnace. The silver-gold residue remains unchanged.
CYCLOTRON
One of the atom-smashing machines used to bombard the nuclei of atoms.

D

DACRON
One of the synthetic fibers.
DDT
Letters stand for dichloro-diphenyl-trichloroethane, an insecticide.
DECOMPOSITION REACTIONS
Those in which a compound is decomposed by heat, light, or electricity into simpler compounds.
DECREPITATION
The expulsion of water with a crackling sound when some crystals are heated.
DEHYDRATION
The removal of water from a substance.
DELIQUESCENCE
The process of picking up enough water to become wet.
DENSITY
Mass per unit volume, e.g., grams per cubic centimeter.
DESTRUCTIVE DISTILLATION
The process of heating wood, coal, bones, etc., in a closed vessel, resulting in a breaking down to simpler materials.
DETERGENT
An agent that removes dirt.
DEUTERIUM
An isotope of hydrogen of mass 2.

DEUTERON
The nucleus of the deuterium atom.

DEVELOPING
A process in photography in which the reduction of the silver compound started by light, is promoted by the action of an alkaline, organic, reducing agent.

DIALYSIS
A process in which a semi-permeable membrane is used to separate colloidal particles from substances in true solution.

DIBASIC
A term applied to acids that have two replaceable hydrogen atoms.

DIESELENE
A petroleum fuel for Diesel engines.

DIFFUSION
The intermingling of liquids and gases.

DILUENT
A diluting agent, such as water in a solution, or turpentine in paint.

DIPOLES
Molecules with unbalanced charges so that one end may be positive and the other end negative.

DISINFECTANTS
Agents that kill, not merely arrest, the growth of bacteria.

DISSOCIATION
The separation of the ions of an electrovalent compound by the action of a solvent.

DISTILLATION
The process of evaporation followed by condensation of the vapors in a separate vessel.

DOUBLE REPLACEMENT REACTIONS
Those in which two compounds exchange ions to produce two new compounds.

DOUBLE SALT
A salt in which two metal atoms are combined with one acid radical or one metal is combined with two acid radicals, e.g., $KAl(SO_4)_2$ 12 H_2O.

DRIER
Catalysts, such as oxides of lead and manganese, added to paint to promote the drying of the paint.

DROSS
A powdery scum that floats on top of melted metal.

DUCTILITY
That property of a substance which permits its being drawn into wire.

DUTCH PROCESS
A process for making white lead using lead buckles, acetic acid, and decomposing tanbark or manure.

DYNAMITE
An explosive made by absorbing nitroglycerin in wood flour mixed with sodium nitrate.

DYNE
A unit of force. The force necessary to give a mass of one gram an acceleration of one centimeter per second per second.

DYNEL
One of the synthetic fibers.

E

EFFERVESCENCE
The process of giving off bubbles of gas from a liquid.

EFFLORESCENCE
The property of giving off water vapor to the air.

ELASTOMER
Actually,it is any flexible or elastic material but,in a more limited sense,a synthetic rubber or soft or rubbery plastic with some degree of elasticity at room temperature.

ELECTRIC POTENTIAL
Electrical pressure or voltage between terminals.

ELECTRODE
A terminal of an electric circuit where the current either enters or leaves.

ELECTROLYTE
A compound whose water solution conducts an electric current.

ELECTRON
A unit particle of negative electricity. Its mass is 1/1845 of the hydrogen atom.

ELECTRON VOLT
That quantity of energy which is equal to the kinetic energy of an electron accelerated by a potential difference of 1 volt.

ELECTRONEGATIVE ELEMENT
An element which has a tendency to take up electrons.

ELECTROPLATING
Deposition of metals on a surface by means of an electric current.

ELECTROTYPES
Copper plates from which the pages of a book are printed.

ELECTROVALENCE
Type of chemical bonding where one or more electrons are transferred from one atom to another.

ELEMENT
Solid,liquid or gaseous matter consisting of atoms of one type which cannot be further decomposed by chemical means. The atoms of an element may differ physically but do not differ chemically. Example: chlorine. Known elements: 101.

EMULSIFIER,EMULSIFYING AGENT
A chemical that mixes and disperses dissimilar materials to produce an emulsion and keep it stable. Casein,for example is a natural emulsifier in milk, keeping butter fat droplets emulsified.

EMULSION
Suspension of insoluble fine particles or globules of a liquid in another liquid.

ENDOTHERMIC
Pertaining to a reaction which absorbs heat.

ENERGY
The capacity for doing work.

ENZYME
An organic secretion that acts as a catalyst.

EQUATION
An expression which shows,by the use of symbols and formulas,the changes in arrangement of the atoms which occur during a chemical reaction.

EQUILIBRIUM
A reaction in which the products unite to form the original reactants at the same speed at which the reactants are forming the products.

EQUILIBRIUM(CHEMICAL)
A state in which a chemical reaction and the reverse reaction are taking place at the same rate. The concentrations (at equilibrium) of all substances remain constant.

EQUILIBRIUM CONSTANT

An equilibrium constant (K) is the ratio (number) obtained by dividing the product of the active concentrations of the substances produced in a reaction by the product of the active concentrations of the reactants, after equilibrium has been reached.

ERG

The work done by a force of one dyne per centimeter.

ESTER

A compound formed by the reaction between an acid and an alcohol.

ESTERIFICATION

The process of preparing esters by adding acid to an alcohol

ETHANE (C_2H_6)

A saturated hydrocarbon (maximum number of hydrogen atoms attached to each carbon)derived from petroleum or natural gas,important for organic synthesis. Ethylene derived from it.

ETHANOL (ETHYL ALCOHOL) (C_2H_5OH)

Organic compound derived through either a fermentation process or via synthesis from petroleum or natural gas. Wide use as solvent and for chemical synthesis.

ETHYLENE (C_2H_4)

Gaseous organic compound prepared in cracking of petroleum or by passing natural gas through heated tube. Removal of two hydrogen atoms from ethane component of petroleum or natural gas makes this unsaturated hydr carbon of wide use as petrochemical base for numerous chemical reactions notably plastic material manufacture.

ETHYLENE GLYCOL

Colorless liquid which is a useful humectant since it absorbs approximately twice its weight of water at room temperature and 100% humidity. A major use: anti-freeze. Among many other uses -chemical synthesis, as in producing alkyd resins.

EUTECTIC

A mixture of two or more substances with the lowest melting point.

EXPLOSIVE RANGE

A pair of percentages below or above which the gas will not form an explosive mixture with air.

EXPOSURE

In photography, the act of admitting light to the film or plate.

EXOTHERMIC

Pertaining to a reaction which liberates heat.

EXTRUDED

Forced through a die by pressure.

F

FAHRENHEIT TEMPERATURE

Temperature on the Fahrenheit scale with 32°as the freezing point of water and 212° as the boiling point at a pressure of 760 mm.

FAMILY OF ELEMENTS

A group of elements with more or less similar properties.

FATS

Glyceryl esters of certain organic acids.

FELDSPARS

Complex silicates, usually aluminum silicate with either sodium or potassium silicate.

FERMENTATION

A chemical reaction caused by living organisms or enzymes.

FERTILIZER

Plant food,or material,that contains compounds of the elements needed by plants for growth.

FILTRATION
The process of removing suspended material from a liquid by allowing the liquid to pass through a material such as filter paper or a layer of sand.

FISSION
The disintegration of an atom into two nyclei with nearly equal mass.

FIXATION OF NITROGEN
A process in which atmospheric nitrogen is converted into useful compounds.

FIXING
In photography the operation of removing unchanged silver salts after the picture has been developed,thereby fixing the image on the film or plate.

FLOTATION (ORE)
A process in which crushed ore is agitated in water containing a frother (pine oil) and a collector (potassium ethyl xanthate). The valuable mineral particles are attached to the froth and rise to the surface from which they are removed.

FLOTATION REAGENT
Chemical used in flotation separation of minerals. Added to pulverized mixture of solids,water and oil,causes preferential oil-wetting of certain solid particles,making possible the flotation and separation of unwet particles.

FLUIDITY
The reciprocal of viscosity.

FLUORESCENT
Giving off light after exposure to sunlight.

FLOWERS OF SULFUR
Finely-divided sulfur formed by the condensation of sulfur vapors on a cool surface.

FLUX
A material added to unite with impurities to form an easily melted product.

FORMULA
A collection of chemical symbols indicating what elements and how many atoms of each are present in a compound.

FORMULA WEIGHT
The sum of the weights of the atoms in a formula.

FOURDRINIER MACHINE
A machine that converts a suspension of taper fibers into sheets.

FRACTIONAL CRYSTALLIZATION
Separation of two dissolved solids by evaporating until the less soluble solid separates while the more soluble solid remains in solution.

FRACTIONATION
The process of separating a mixture by careful evaporation,depending upon the materials having different boiling points.

FROTH-FLOTATION
A process for concentrating powdered ore by causing the good ore to cling to bubbles which float above liquid,while the worthless rocky material sinks to the bottom of the container.

FUNGICIDE
Any one of a group of chemicals used to prevent or inhibit the growth of fungi or bacteria. Among these are: plant fungicides,wood preservatives,mildew or mold preventives and disinfectants.

G

GALVANIZE
To coat iron or steel with zinc.
GAMMA RAYS
High energy X-rays emitted from a radioactive material.
GANGUE
Worthless material, rock or earth, present in an ore.
GEIGER COUNTER
A device used to detect the presence of radiation from radioactive material.
GEL
A jelly-like solid.
GENERATOR
In chemistry, the vessel in which a reaction occurs between chemicals.
GERMAN SILVER
An alloy containing copper,nickel, and zinc.
GRAM
Basic unit of weight in the metric system,equal to 1/1000 of the standard kilogram.
GRAM-EQUIVALENT WEIGHT
The weight of an element that will combine with or replace one gram of hydrogen.
GRAM-FORMULA WEIGHT
The number of grams of a substance that equal its formula weight.
GRAM-MOLECULAR VOLUME
The volume,22.4 liters, occupied by one gram-molecular weight at any gast at S.T.P.
GRAM-MOLECULAR WEIGHT
That number of grams of any substance that equal its molecular weight, also called a MOLE of that substance

H

HALF-LIFE
The time required for one-half the atoms of a mass of radioactive material to decompose.
HALOGEN
The name given to the family of elements having seven valence electrons.
HEAT OF FORMATION
The heat which is given out or absorbed when a compound is formed from elements.
HEAT OF NEUTRALIZATION
The number of calories liberated in the formation of 18g. of water from hydrogen and hydroxyl ions.
HEAVY WATER
Water containing deuterium atoms in place of ordinary hydrogen atoms.
HEMOGLOBIN
The red coloring matter in blood.
HERBICIDE
A weed-killing agent. Most of these are specific in their action and therefore not intended for indiscriminate use. Label indicates particular purposes and gives directions for most effective application.
HOMOLOGOUS SERIES
A series of compounds each of which can be represented by a type of formula, such as $C_{\eta}H_{2\eta+2}$.
HUMIDITY,RELATIVE
The ratio of the actual amount of water vapor in atmosphere to the amount necessary for saturation at the same temperature.

HYDRATE
Crystals that contain water of hydration.

HYDRATED LIME
Calcium hydroxide,the product formed when water unites with calcium oxide.

HYDROCARBONS
Organic compounds composed solely of caron and hydrogen. Myriad variety of molecular combinations of C and H. Basic building block of all organic chemicals. Main chemical industry course of hydrocarbons: petroleum, natural gas and coal.

HYDROFORMING
The process of improving gasoline by heating it with hydrogen in the presence of a catalyst.

HYDROGENATION
The addition of hydrogen to a material.

HYDROLYSIS
The reaction of water with a salt to form the acid and the base of which the salt was a product; it opposes neutralization reactions.

HYDROPONICS
The science of gardening without the use of oil.

HYGROSCOPIC
Having a tendency to pick up water vapor.

I

INACTIVE SUBSTANCE
One which reacts, but not very readily, with other substances.

INDICATOR
A substance used to show,by means of a color change,whether an acid or a base is present.

INERT ELEMENT
An element of the zero group of the periodic table. Elements in this group have no chemical properties.

INERT SUBSTANCE
One which does not react at all with other substances under the usual conditions of chemical reactions.

INHIBITOR
(1) A material used to prevent or retard rust or corrosion.
(2) An agent which arrests or slows chemical action.

INORGANIC
Term used to designate chemicals that generally do not contain carbon. Source: matter, other than vegetable or animal. Example: chlorine is an inorganic chemical derived from salt.

INSECTICIDE
Any one of a group of chemicals used to kill or control insects.

INTERMEDIATE
An organic chemical formed as a "middle-step" between the initial material and the one or frequently several ultimate end products.

INVERSION
The combination of cane sugar with water to form two molecules of simple sugars.

ION
An atom or group of atoms which carries an electric charge.

ION-EXCHANGE RESINS
Granules of resins that absorb either positive or negative ions.

IONIC EQUILIBRIUM
The balance attained when the rate of dissociation equals the rate of association.

IONIZATION
The formation of ions from polar compounds by action of a solvent.
IONIZATION CONSTANT
The product of the concentration of the ions divided by the concentration of the unionized molecules of solute (electrolyte).
IONIZATION POTENTIAL
The energy necessary to remove an electron from a gaseous atom to form an ion. This energy is expressed in electron volts.
IRRADIATION
Subjected to light, especially ultraviolet light.
ISOBARES
Atoms of the same atomic weight but having different atomic numbers are isobares.
ISOMERS
Compounds having the same composition but different structure.
ISOTOPE
One of two or more atomic species of an element differing in weight but having the same nuclear charge (atomic number). For example, in the element, chlorine, the atomic weight is the mean of the two isotopes making up the element.

K

KAOLIN
A fine, white clay composed of hydrated aluminum silicate.
KERNEL
All of the atom except the valence electrons.
KILN
A type of furnace used for producing quicklime, making glass, baking pottery, etc.
KILOCALORIES
Units equal to one thousand calories.
KILOGRAM
The standard of weight in the metric system equal to 1000 grams.
KINDLING TEMPERATURE
The lowest temperature at which a substance takes fire. This temperature varies with the physical state of the substance.
KINETIC ENERGY
Energy of motion.
KINETIC THEORY
The theory of matter which assumes that all molecules of matter are always in motion.
KNOCKING
A pounding sound produced in automobile engines by too rapid combustion of the mixture of gasoline vapor and air.

L

LAC OF SULPHUR
Precipitated sulfur.
LATENT HEAT
The heat absorbed or liberated in changing a mole of substance from one state to another at a fixed temperature, e.g., converting 18g. water to water vapor at 100°C.
LATEX
Original meaning: Milky extract from rubber tree. Now also applied to water emulsions of synthetic rubbers or resins. In emulsion paints, the film-forming resin is in the form of latex.
LAW, LeCHATELIER'S
A system in equilibrium, if disturbed by external factors such as temperature and pressure, will adjust itself in such a way that the effect of the disturbing factors will be reduced to a minimum.

LEHR
A cooling oven for annealing glass.

LIGNIN
Major non-carbohydrate constituent of wood and woody plants; functions as binder for the cellulose fibers. Removed from wood in pulp manufacture. Extracted from waste sulfite liquor. Research underway on chemical applications. Current use as adhesive base, for boiler water treatment and for road binders.

LIGNITE
A partially mineralized peat.

LIME
A term loosely used for all calcium compounds but properly belonging to calcium oxide, although often used for calcium hydroxide and calcium carbonate.

L.P.G., or LIQUEFIED PETROLEUM GAS
A compressed or liquefied gas comprised of pure propane, or a butane, or a combination of propane and butane; obtained as a by-product in petroleum refining or gasoline manufacture. Used in chemical synthesis.

LITER
The basic unit of volume in the metric system.

LITHOPANE
A paint base composed of barium sulfate and zinc sulfide.

LITMUS
A dye extracted from lichens which is used as an indicator.

LYE
A term used for either sodium hydroxide or potassium hydroxide.

M

MASS
The property of a substance(body) that determines the acceleration it will acquire when acted upon by a given force.

MASS ACTION, LAW OF
The speed of a chemical change is proportional to the concentration of the reacting substances.

MATTE
A mixture of sulfides produced in a partially refined ore.

MATTER
Anything which occupies space and has weight.

METALLURGY
The science of extracting and refining metals.

METALS
Elements with a luster that are good conductors of heat and electricity and are electro-positive.

MERCERIZING
Treating stretched cotton with sodium hydroxide solution.

METAMORPHIC
Applied to rocks that have undergone a change in form due to heat or pressure.

METER
The basic unit of the metric system, equal to 39.37 inches.

METHANE (CH_4)
The simplest saturated hydrocarbon, chief component of most natural gas. Chemical raw material.

METHYL ALCOHOL (METHANOL, WOOD ALCOHOL) (CH_3OH)
Organic compound important for chemical synthesis; also used in denaturing alcohol; solvent and many other uses.

MILLILITER
One-thousandth of a liter.
MILK OF SULFUR
Precipitated sulfur.
MIXTURE
Two or more substances which, when combined, do not lose their identity and may be separated by mechanical means.
MODERATOR
A substance which slows down fast neutrons.
MOLAL SOLUTION
One mole of a substance dissolved in 1000g. of solvent.
MOLAR SOLUTION
One gram-molecular weight of a substance dissolved in enough solvent to make one liter of solution.
MOLE
That number of grams of a substance which is exactly equal to its molecular weight.
MOLECULAR VOLUME
The volume occupied by a mole of any gas at 0°C. and 760 mm pressure, e.g., 22.4 liters.
MOLECULAR WEIGHT
The sum of the weights of the atoms in a molecule.
MOLECULE
The chemical combination of two or more like or unlike atoms.
MONATOMIC
Molecules made up of one atom.
MONOBASIC ACID
An acid having one replaceable hydrogen atom per molecule.
MONOCLINIC
Referring to those crystals having one oblique axis.
MONOMER
A compound of relatively low molecular weight which, under certain conditions, either alone or with another monomer, forms various types and lengths of molecular chains called polymers or copolymers of high molecular weight. Example: styrene is a monomer which polymerizes readily to make the polymer, polystyrene.
MORTAR
A mixture of lime, sand, and water.
MOTHER LIQUOR
The liquid which is left after a crop of crystals has separated from a solution.

N

NAPHTHALENE ($C_{10}H_8$)
A white solid crystalline hydrocarbon found as a mineral and obtained from coal tar by distillation. Used as a moth repellent and a basic material in the manufacture of dyestuffs, synthetic resins, lubricants and other products.
NAPHTHENES
Hydrocarbons having a ring structure.
NASCENT
At the instant an element is liberated from a compound it is said to be in the nascent state. Nascent dryogen is probably atomic hydrogen.
NATIVE METAL
A metal found as an element, rather than as a compound, in the ground.

NATURAL GAS
A combustible gas composed largely of methane and other hydrocarbons with variable amounts of nitrogen and non-combustible gases; obtained from natural earth fissures or from driven wells. Among other things, used as a fuel, in the manufacture of carbon black and in chemical synthesis of many products.

NEON (Ne)
A rare gaseous element which forms no chemical compounds and is derived by fractional distillation of liquid air. Used mostly in luminescent electric tubes.

NEUTRALIZATION
The union of hydrogen ions of an acid with hydroxyl ions of a base to form water.

NEUTRON
A neutral particle found in the atom.

NIACIN ($C_6H_5O_2N$)
The anti-pellagra factor of the vitamin B complex; present in animal tissues, in fish, milk and green leafy vegetables.

NITER (KNO_3)
A white salt widely distributed in nature and formed in soils from nitrogenous organic bodies by the action of bacteria. Used in making gunpowder, medicinals and other products.

NITRIC ACID (HNO_3)
A colorless to yellowish fuming liquid with powerful corrosive properties. Manufactured by several methods, it is used in organic synthesis, in etching metals and ore flotation, in the manufacture of explosives, medicines, and other products.

NITROCELLULOSE
A powerful explosive made by treating cellulose with nitric and sulfuric acids.

NITROGATION
Adding of nitrogen compounds to the soil.

NITROGEN CYCLE
The cycle of changes through which nitrogen passes, starting with nitrates in the soil which become, in turn, plant proteins, animal proteins, dead matter, ammonia, nitrites, and then nitrates again.

NITROGEN FIXATION
Process of combining nitrogen of the atmosphere into any of the various stable chemical compounds valuable to the manufacture of fertilizers, and ammonia, among others.

NITROGLYCERIN
Glyceryl trinitrate, a powerful and sensitive explosive.

NITROUS OXIDE (NO_2)
A colorless gas of sweetish odor and taste; used as an anesthetic. Also called laughing gas.

NONELECTROLYTE
A compound whose water solution does not conduct electric current.

NONMETALS
Elements that are usually poor conductors of heat and electricity and are electro-negative.

NORMAL SALT
A salt containing neither replaceable hydrogen nor hydroxyl.

NORMAL SOLUTION
A solution that contains one gram-equivalent of a substance dissolved in enough solvent to make one liter of solution.

NUCLEONS
The fundamental constituents of atomic nuclei (protons and neutrons).

O

OCCLUSION
The adsorption of gases by solids.

OCTANE RATING
A number indicating how a gasoline behaves with regard to knocking when compared with a test fuel given an arbitrary rating of 100.

OCTET
The term applied to a group of eight electrons in the highest energy levels of atoms.

ONE ATMOSPHERE
The average pressure of the atmosphere at sea level, equal to 760 mm. of mercury.

OPEN-HEARTH PROCESS
A process for making steel in a large, shallow pool.

ORBIT
The path of an electron about the nucleus of an atom.

ORGANIC
Term used to designate that group of chemicals that contain carbon. Approximately 300,000 such compounds have been identified, many occurring in nature, others produced by chemical synthesis. Sources: petroleum and coal tar by fractional distillation; soft coal and wood by destructive distillation; wood by chemical treatment; grains and fruits by fermentation; grains, vegetables and fruit by mechanical and chemical separation of starch and sugar; cotton by mechanical and chemical treatment; animals, seeds and nuts, by mechanical extraction of fats and oils.

ORLON
One of the synthetic fibers.

OSMOSIS
The passage of liquids and gases through porous membranes.

OXIDATION
Process of combining oxygen with some other substance.

OXYGEN-CARBON DIOXIDE CYCLE
The cycle of events whereby plants take in carbon dioxide and give off oxygen in photosynthesis, whereas animals take in oxygen and give off carbon dioxide in respiration.

OZONE
An allotropic form of oxygen containing three atoms per molecule.

P

PAINT BASE
The particles suspended in the oil of a paint.

PAINT VEHICLE
A quick drying oil that forms a flexible horn-like film. The paint base is suspended in this oil.

PARAFFINS, PARAFFIN SERIES (From parun affinis-small affinity)
Those hydrocarbon components of crude oil and natural gas whose molecules are saturated (i.e., carbon atoms attached to each other by single bonds) and therefore very stable. Examples: methane, ethane.

PARKERIZED
Dipped into a hot alkaline solution of sodium phosphate.

PARKES PROCESS
A method of separating silver from molten, crude lead by adding zinc.

PEPTIZATION
A breaking up of coarse particles into a finer state of subdivision.

PERIODIC TABLE
An arrangement of the elements in the order of their atomic numbers.

pH
A numerical scale that indicates the concentration of hydrogen ions in a solution; 7 is neutral, less than 7 is acid, and greater than 7 is basic.

PHENOL (C_6H_5OH)

Popularly known as carbolic acid. Important chemical intermediate intermediate derived primarily from coal tar and produced by chemical synthesis. Base for plastics,pharmaceuticals,explosives,anti-septics, and many other end products.

PHENOLPHTHALEIN

An indicator which turns red in the presence of an excess of hydroxyl ions.

PHOSPHORS

Compounds that fluoresce under ultraviolet light.

PHOSPHORESCENCE

A faint glow similar in appearance to that emitted by phosphorus when it is exposed to the air in a dark room.

PHOTON

A unit of light (a particle of light).

PHOTOSYNTHESIS

The process by which plants build carbohydrate foods with the aid of sunlight,using carbon dioxide and water as the raw materials and chlorophyll as the catalyst.

PHYSICAL CHANGE

A change in color, size of particle, temperature, or other physical property that does not produce a new substance.

PHYSICAL EQUILIBRIUM

A condition of balance when the rate of a physical change in one direction is equal to an opposite physical change.

PICKLING

Treating a metal with acid to remove surface coatings of oxide.

PIGMENT

A substance that adds color to a mixture.

PLASTICIZERS

Organic chemicals used in modifying plastics,synthetic rubber and similar materials to give such special properties as elongation, flexibility and toughness as may be essential to their end uses.

PLASTICS

Officially defined as any one of a large and varied group of materials which consists of, or contains as an essential ingredient,an organic substance of large molecular weight; and which,while solid in the finished state,at some stage in its manufacture has been or can be formed (cast calendered,extruded,molded,etc.) into various shapes by flow - usually through application of heat and pressure singly or together. Each plastic has individual physical,chemical and electrical properties. Two basic types: thermosetting and thermoplastic.Prior to processing,plastic materials often are referred to as resins. Final form may be as film,sheet,solid,or foam; flexible or rigid.

POLAR MOLECULE

A molecule with an unsymmetrical electron distribution.

POLING

The use of green wood in the refining of copper ore to reduce the traces of copper oxide present.

POLYMER

A high molecular weight material containing a large number of repeating units. These may be hundreds or even thousands of the original molecules (monomers)which have linked together end to end. Rubber and cellulose are naturally occurring polymers. Most resins are chemically produced polymers. Polymers may be formed by polymerization or condensation. For instance,the polymer,polyethylene, is polymerized from the monomer, ethylene. An example of condensation is the production of phenol formaldehyde resins with the incidental formation of water or some simple substance.

POLYMERIZATION
A physical reaction by which polymers are formed from the linkage of monomers.
POLYMORPHISM
The ability to exist in two or more crystalline forms.
PORCELAIN
A product made from pure, white clay mixed with powdered feldspar and usually fired twice in a kiln.
POSITRON
A unit charge of positive electricity of approximately the same mass as the electron.
POTASH
Source of potassium, essential plant nutrient (other two basic nutrients: nitrogen, phosphorus). Potash value in a fertilizer is expressed in terms of equivalent amount of potassium oxide K_2O.
PRECIPITATE
An insoluble solid formed by adding one solution to another.
PRODUCER GAS
A cheap fuel gas for industrial purposes made by blowing a blast of steam and air through red-hot coke.
PROPERTIES
The characteristics by which we identify materials.
PROTEINS
Complex organic compounds necessary for the growth of living things or the repair of worn-out tissue.
PROTON
A positively charged particle found in the atom.

Q

QUICKLIME
Calcium oxide, often called unslaked lime.

R

RADICAL
A group of atoms which acts like a single atom in forming compounds.
RADIOACTIVITY
Emission of energy in waves or moving particles from the nucleus of an atom. Always involves change of one kind of atom into a different kind. A few elements such as radium are naturally radioactive. Other radiactive forms are induced (see RADIOISOTOPE).
RADIOCHEMICALS
Any compound or mixture containing a sufficient portion of radioactive elements to be detected by Geiger counter.
RADIOISOTOPE
An isotopic form of an element that exhibits radioactivity,whether naturally found or produced by fission and other induced nuclear changes. The latter are used in biological tracer work and industrial control operations. More than 500 radioactive substances have been produced.
RADON
A gaseous element produced by the disintegration of radium atoms.
REACTANTS
The elements or compounds entering into a chemical reaction.
REAGENT
Any substance used in a chemical reaction to produce another substance or to detect its composition.
REDOX REACTION
A reaction in which oxidation and reduction take place.
REDUCING AGENT
An agent that removes oxygen from a material.

REDUCTION
(1) Removal of oxygen. (2) The gain in electrons by an element.

REFRACTORY
Pertaining to a substance that is not easily melted.

REGENERATIVE HEATING
A system whereby the heat of outgoing flue gases is used to preheat incoming fuel gas and air.

RELATIVE HUMIDITY
The amount of moisture present in the air as a vapor compared with the total amount of moisture the air could hold at that temperature.

REPLACEMENT SERIES
The arrangement of the metals in the order of their decreasing chemical activity.

RESIN
A solid or semisolid amorphous organic compound or mixture of such compounds with no definite melting point and no tendency to crystallize. Resins may be of vegetable, animal or synthetic origin. Natural resins may be distinguished from gums in that they are insoluble in water. However, certain synthetic water soluble materials are referred to as resins or resin stages. There are many types, each with distinctive physical and chemical properties. Some types of resin materials may be molded, cast, or extruded. Others are used for adhesives, for the treatment of textiles and paper, and for protective coatings. Still others are rolled or extruded into continuous sheets and films of various thicknesses. All, broadly speaking, have plastic use.

RHOMBIC
Referring to crystals having equilateral edges and oblique angles.

RIFFLES
Grooves in a sluice for catching gold in hydraulic mining operations.

S

SACRIFICIAL METAL
The more active of a pair of metals which is oxidized by an electric cell action.

SALTS
Crystalline compounds made up of metals and non-metals. Example: table salt is sodium and chlorine.

SAPONIFICATION
The process of making soap by adding lye to a glyceryl ester.

SARAN
One of the synthetic fibers.

SATURATED SOLUTION
A solution in which the solute in solution is in equilibrium with undissolved solute.

SEDIMENTARY
A term applied to rocks formed from sediment that has been deposited in layers.

SHELL-FILLER
An explosive that requires a severe shock to set it off.

SHERARDIZING
Coating with zinc by allowing zinc vapor to condense on the object.

SILICONES
Unique new group of polymers made by molecular combination of the inorganic chemical, silicone dioxide, with organic chemicals. Produced in variety of forms including silicone fluids, resins and rubber. Silicones have special properties, such as water-repellency, wide temperature resistance, durability and dielectric property

SILT
A soil which is coarser than clay but finer than sand.
SINGLE REPLACEMENT REACTION
One in which an element replaces another in a compound.
SINTERING
Heating until fusion just begins.
SLAG
A by-product of smelting. It is formed by the action of low melting material (flux) on impurities in the ore. Slags contain calcium and aluminum silicates.
SLAKED LIME
Calcium hydroxide, or calcium oxide, that has united with water.
SMOKELESS POWDER
A nitrocellulose explosive.
SOAP
The sodium or potassium salts of a fatty acid, e.g., $NaC_{18}H_{35}O_2$.
SOLDER
An easily melted alloy, especially one of tin and lead.
SOLUBILITY PRODUCT CONSTANT
The product of the concentrations of the ions of slightly soluble salt at saturation.
SOLUTION
Mixture in which the identities of components are lost as such.
SOLVENT
A substance, usually an organic compound, which dissolves another substance.
SPECIFIC GRAVITY (GASES)
The ratio of the weight of one liter of air to the weight of one liter of the gas.
SPECIFIC GRAVITY (SOLID OR LIQUID)
The ratio of the weight of a unit volume of a substance to the weight of the same volume of water.
SPECIFIC HEAT
The heat required to raise the temperature of one gram of a substance one degree centigrade.
SPECIFIC VOLUME
The volume of one gram of a substance.
SPINNERETS
Thimble-like plates with tiny holes for extruding synthetic fibers.
STALACTITES
Icicle-like masses of calcium carbonate hanging from the roof of limestone caves.
STALAGMITES
Masses of calcium carbonate rising from the floor of limestonn caves, formed by dripping of calcium bicarbonate solution from the cave roof.
STANDARD CONDITIONS
0°C. and 1 atmosphere pressure (760 mm.).
STANDARD PRESSURE
A pressure equal to that furnished by a column of mercury 760 mm. high.
STANDARD TEMPERATURE
Zero degrees Centigrade.
STERLING
Containing 92.5% silver, or 925 fine.
S.T.P.
Abbreviation for "standard temperature and pressure."
STRONG ACID
One which is completely ionized in water solutions.

STRONG ELECTROLYTE
One which is ionized almost completely.
STRUCTURE FORMULAS
Formulas that tell how the atoms are joined in the molecules.
SUBLIMATION
A process in which a solid is vaporized and condensed to a solid without passing through the liquid state.
SUPERHEATED WATER
Water heated under pressure to a temperature above the normal boiling point.
SUPERPHOSPHATE
Phosphorus-bearing material made by action or sulfuric acid on phosphate rock to make phosphorus available as a plant nutrient. Phosphorus in soluble form is one of the three essential plant nutrients. (Other two: nitrogen, potassium.)
SUPERSATURATED
Pertaining to a solution saturated at a high temperature which retains the solute in solution as it cools.
SURFACE-ACTIVE AGENT
Any of a group of compounds added to a liquid to modify surface or interface tension. In the case of synthetic detergents, best known surface-active agent,reduction of tension provides cleansing action. Term also includes dispersing, emulsifying,foaming,penetrating and wetting agents. Usually synthetic organic in origin.
SURFACE COATINGS
Term used to cover paint,lacquer,varnish and other chemical compositions used for protecting and/or decorating surfaces.
SURFACTANT
A coined word which means "surface-active agent."
SYNCHROTRON
One of the "atom-smashing" machines for bombarding atomic nuclei.
SYNERGIST
A material which,in combination with another,improves the effectiveness of the combination to a degree in excess of the sum of the effects of the two materials taken independently.
SYNTHESIS
The reaction,or series of reactions,by which a complex compound is obtained from simpler compounds or elements.
SYNTHETIC DETERGENTS
Chemically-tailored cleaning agents soluble in water. Originally developed as soap substitute. Because they do not form insoluble precipitates,they are especially valuable in hard water. See SURFACE-ACTIVE AGENT.
SYNTHETIC RUBBER
Nab-made polymeric chemical with rubber-like attributes. Various types with varying composition and properties. Major types designated as S-type,butyl,neoprene (chloroprene polymers), and N-type.

T

TALL OIL (Name derived from Swedish TALLOLJA;material first investigated in Sweden-not synonymous with our pine oil.)
Natural mixture of rosin acids,fatty acids,sterols,high molecular weight alcohols and other materials,derived primarily from waste liquors of wood pulp manufacture. Dark brown,viscous,oily liquid often called liquid rosin. Recent encyclopedia lists 38 major industry uses.
TANKAGE
A fertilizer made from slaughter house scraps.

TAR CRUDE
Basic organic raw material derived from distillation of coal tar and used for chemical manufacture.

TEMPERING
Regulation of the iron carbide, and thus the hardness of a piece of steel, by heating and sudden cooling.

TERNARY
Composed of three elements.

TERNE PLATE
Sheet iron coated with an alloy of tin and lead.

THERMITE REACTION
The replacement reaction between a metal, such as aluminum, and an oxide, such as ferric oxide, which liberates much heat.

THERMOPLASTIC TYPE OF PLASTICS
Those that can repeatedly melt or soften with heat and harden on cooling. Examples: vinyls, acrylics, polyethylene.

THERMOSETTING TYPE OF PLASTICS
Those that are heat-set in their final processing to a permanently hard state. Examples: phenolics, ureas and melamines.

THINNER
A liquid, such as turpentine, used in paint to make it spread more easily and penetrate better

TINCTURE
A solution the solvent of which is alcohol, e.g., tincture of iodine.

TITRATION
Determination of the concentration of a solution by comparing it with a standard solution, usually employing burettes for the operation.

TOLUENE ($CH_3C_6H_5$)
Hydrocarbon derived mainly from petroleum but also from coal. Base for TNT, lacquers, saccharin and many other chemicals.

TRIADS
Group of three elements in the same vertical column of the periodic table.

TRIBASIC ACID
An acid containing three replaceable hydrogen atoms per molecule.

TRIPLE SUPERPHOSPHATE
Phosphorus fertilizer of higher phosphorus content than in superphosphate; produced by addition of phosphoric acid to phosphate rock.

TUYERES
The nozzles of the blowpipes of a blast furnace.

TYNDALL EFFECT
The dispersion of a beam of light as it passes through a colloidal solution.

U

ULTRAVIOLET RAYS
Invisible radiations having a shorter wave length than violet light.

UNSLAKED LIME
Calcium oxide, or line to which water has not been added.

U.S.P.
Abbreviation for United States Pharmacopoeia, an official book that specifies strengths and degrees of purity of official remedies.

V

VALENCE
The number of electrons gained, lost, or shared in forming a chemical bond.

VALENCE ELECTRONS
Electrons in the outermost orbit which may be gained or lost in chemical reactions.

VAPOR
A gas that can be converted into a liquid at that temperature by pressure alone.

VAPOR DENSITY
The ratio of the weight of a gas to the weight of an equal volume of hydrogen measured under the same conditions.

VAPOR PRESSURE
The (partial) pressure exerted by a vapor.

VAPOR TENSION
The pressure which a vapor exerts on a liquid when the liquid and vapor are in equilibrium at a given temperature (the maximum vapor pressure for the given temperature).

VAT DYES
Water-insoluble,complex coal tar dyes that can be chemically reduced in a heated solution to a soluble form that will impregnate fibers. Subsequent oxidation then produces insoluble colored dye-stuff remarkably fast to washing, light and chemicals.

VEHICLE
The liquid portion of paint,such as linseed oil,used to hold the paint base in suspension.

VERMICULITE
Mica that has been expanded with steam to make a light,porous material.

VISCOSITY
(1) The resistance to flow of a liquid.
(2) The internal friction of a liquid.

VITRIFIED
Heated to the point where melting just begins,thus closing pores.

VITRIOL
An acid substance.

VOLUME-VOLUME PROBLEMS
Those in which a known volume of one material is given and the volume of another material involved in the reaction is sought.

VOLT
The electrical pressure required to make a current of one ampere through a resistance of one ohm.

VULCANIZATION
Process of combining rubber (natural,synthetic,or latex) with sulfur and various other additives usually under heat and pressure,in order to eliminate tackiness when warm and brittleness when cool,and to change permanently the material from a thermoplastic to a thermosetting composition. Finally to otherwise improve strength,elasticity, and abrasive resistance.

W

WARFARIN
A new kind of rat poison.

WATER GAS
A fuel gas made by blowing a bast of steam through a bed of red-hot coke.

WATER GLASS
A syrupy solution of sodium silicate.

WATER OF HYDRATION
Water that has united with some chemicals as they form crystals called hydrates.
WEAK ACID
An acid which is but slightly ionized in water solutions.
WEAK ELECTROLYTE
One that is but slightly ionized.
WEIGHTED SILK
Silk that has been dipped in solutions of certain tin salts.
WEIGHT-WEIGHT PROBLEMS
Those in which a known quantity of one material is given and the amount of another material involved in the reaction is sought.
WEIGHT-VOLUME PROBLEMS
Those in which the weight or volume of one material is given and either the weight or volume of another material involved in the reaction is sought
WETTING AGENTS
Materials that reduce the surface tension of a liquid, causing it to spread out better.

X

X-RAYS
Light radiations of high frequency and very short wave length.

Z

ZEOLITES
Naturally occurring minerals, such as an aluminate of either sodium or potassium.
ZIRCONIUM
A metallic element with an exceedingly high melting point.

CHEMICAL LAWS

1. AVOGADRO'S HYPOTHESIS
Equal volumes of gases, under the same conditions of temperature and pressure, contain the same number of molecules.
A liter of oxygen contains the same number of molecules as a liter of hydrogen measured under the same conditions of temperature and pressure.
2. BOYLE'S LAW
If the temperature remains constant, the volume of a given mass of gas is inversely proportional to the pressure.

$$\frac{V_1}{V_2} = \frac{P_2}{P_1}$$

Example: If the pressure on a gas is doubled and the temperature is constant, the new volume is one-half the original volume.
3. CHARLES' LAW
If the pressure remains constant, the volume of a gas caries directly with the Absolute temperature.

$$\frac{V_1}{V_2} \quad \frac{T_1}{T_2}$$

Example: If the Absolute temperature of a gas is doubled, the volume is also doubled.

4. GAY-LUSSAC'S LAW OF COMBINING VOLUMES OF GASES

The relative combining volumes of gases and their products, if gaseous, may be expressed by small whole numbers.

Example: One volume of chlorine combines with one volume of hydrogen forming two volumes of hydrogen chloride.

5. HENRY'S LAW OF GAS SOLUBILITY

If the temperature is held constant, the weight of a gas which dissolves in a given volume is proportional to the pressure.

Example: The greater the pressure, the larger the volume of gas which can be dissolved. Suppose that one liter of water can dissolve 2 grams of carbon dioxide at 1 atmosphere of pressure. If the temperature remains the same, 4 grams can be dissolved if the pressure is increased to 2 atmospheres.

6. LAW OF CONSERVATION OF ENERGY

Energy can neither be created nor destroyed; it may, however, be changed from one form to another.

Example: When coal is burned, stored chemical energy is converted and released as heat energy.

7. LAW OF CONSERVATION OF MATTER (MASS)

Matter can neither be created nor destroyed. If the matter is changed from one form to another, the new products produced have the same mass as the original substances.

8. LAW OF MASS ENERGY

This law expresses the equivalence of mass and energy. It states that energy (E) is equal to mass (m) times the square of the velocity of light (c) in centimeters per second.

$$E = mc^2$$

9. LAW OF DEFINITE PROPORTIONS

Every chemical compound has a definite composition of weight.

Example: If pure water is analyzed its composition will never vary. There will always be eight times as much oxygen by weight as hydrogen.

10. LAW OF MASS ACTION

The speed of a chemical reaction is proportional to the product of the molecular concentration of the reacting substances.

Example: Mass action refers to the changing of the equilibrium of a reaction by varying the concentration of one or more of the reactants. Thus, the law states that in a reversible reaction the speed at which it occurs depends on the concentration of the reactants.

11. LAW OF MULTIPLE PROPORTIONS

When two elements unite to form more than one compound, the weights of one element which combine with a fixed weight of the other are in the ratio of small whole numbers.

Example: In the compound carbon monoxide (CO), there are 12 grams of carbon and 16 grams of oxygen. In the compound carbon dioxide (CO_2), there are 12 grams of carbon and 16 grams of oxygen. Therefore,

$$\frac{\text{weight of oxygen in CO}}{\text{weight of oxygen in } CO_2} = \frac{1}{2}$$

12. LE CHATELIER'S PRINCIPLE

If a system which is in equilibrium is affected by a change in temperature or pressure, it will adjust itself so that the effect of the change will be reduced to a minimum.

Example: If the pressure on a system in equilibrium is increased, the system will adjust itself so that it will occupy less volume.

13. PERIODIC LAW (MOSELEY'S)

The chemical properties of the elements are a periodic function of their atomic numbers.

GLOSSARY OF COMMON SUBSTANCE

COMMON NAME	CHEMICAL NAME	FORMULA
	A	
Alum	Potassium aluminum sulfate	$K_2SO_4.Al_2(SO_4)_3.24H_2O$
Alumina	Aluminum oxide	Al_2O_3
Alundum	Fused aluminum oxide	Al_2O_3
Ammonia water	Ammonium hydroxide	NH_4OH
Aniline	Phenyl amine	$C_6H_5NH_2$
Aqua ammonia	A solution of NH_3 in H_2O	NH_4OH
Aqua fortis	Nitric acid	HNO_3
Aqua regia	Hydrochloric and nitric acids	$3\ HCl + HNO_3$
Asbestos (principal form)	Hydrated magnesium silicate	$3\ MgO\ .\ 2\ SiO_2.2H_2O$
Aspirin	Acetylsalicylic acid	$C_6H_4(COOH)OCOCH_3$
	B	
Babbitt	Alloy of Sn, Sb, and Cu	
Bakelite	Resin from phenol and formaldehyde	
Baking powder	A mixture of $NaHCO_3$, an acid ingredient, and starch	
Baking soda	Sodium bicarbonate	NaHCO
Bauxite	Hydrated aluminum oxide	$Al_2O_3.\ 3H_2O$ and $Al_2O_3.\ H_2O$
Bleaching powder	Calcium oxychloride	$CaOCl_2$
Blue vitriol	Cupric sulfate	$CuSO_4$
Bluestone	Cupric sulfate	$CuSO_4$
Bone ash	Calcium phosphate (impure)	$Ca_3(PO_4)_2$
Bone black	Animal charcoal	C
Boracic acid	Boric acid	H_3BO_3
Borax	Sodium tetraborate	$Na_2B_4O_7.10\ H_2O$
Brass	Alloy of Cu and Zn	
Brimstone	Volcanic sulfur	Impure S
Brine	Sodium chloride solution	NaCl and H_2O
Bronze	Alloy of Cu, Sn, and Zn	
	C	
Cadium yellow	Cadmium sulfide	CdS
Calcite	Calcium carbonate	$CaCO_3$
Calich	Sodium nitrate(impure)	$NaNO_3$
Calomel	Mercurous chloride	HgCl
Camphor, artificial	Pinene chloride	$C_{10}H_{17}Cl$
Cane sugar	Sucrose	$C_{12}H_{22}O_{11}$
Carbolic acid	Phenol	C_6H_5OH
Carbonic acid gas	Carbon dioxide	CO_2

GLOSSARY OF COMMON SUBSTANCES (CONT'D)

COMMON NAME	CHEMICAL NAME	FORMULA
Caustic potash	Potassium hydroxide	KOH
Caustic soda	Sodium hydroxide	$NaOH$
Ceruse	Basic lead carbonate	$2\ PbCo_3.\ Pb(OH)_2$
Chalk	Calcium carbonate (impure)	$CaCO_3$
Chile saltpeter	Sodium nitrate (impure)	$NaNO_3$
China clay	Aluminum silicate	$Al_2O_3.2SiO_2.\ 2H_2O$
Chloride of lime	Calcium oxychloride	$CaOCl_2$
Chrome alum	Potassium chromium sulfate	$K_2SO_4.Cr_2(SO_4)_3.24\ H_2O$
Cinnabar	Mercuric sulfide	HgS
Coke	Carbon (impure)	C
Common salt	Sodium chloride	$NaCl$
Copperas	Ferrous sulfate	$FeSO_4.\ 7H_2O$
Corrosive sublimate	Mercuric chloride	$HgCl_2$
Corundum	Aluminum oxide	Al_2O_3
Cream of tartar	Potassium hydrogen tartrate	$KHC_4H_4O_6$
	D	
Dextrose	Glucose	$C_6H_{12}O_6$
Diamond	Carbon	C
Dry ice	Solid carbon dioxide	CO_2
	E	
Epsom salts	Magnesium sulfate	$MgSO_4$
	F	
Feldspar (one form)	Potassium aluminum silicate	$KalSi_3O_8$
Firedamp	Methane	CH_4
Flowers of sulfur	Sulfur	S
Fluorspar	Calcium fluoride	CaF_2
Fool's gold	Iron pyrite	FeS_2
	G	
Galena	Lead sulfide	PbS
Glass	A solid solution containing a mixture of silicates	
Glauber's salt	Sodium sulfate	$Na_2SO_4.\ 10H_2O$
Glucose	Dextrose	$C_6H_{12}O_6$
Glycerin	Glycerol	$C_3H_5(OH)_3$
Grain alcohol	Ethyl alcohol	C_2H_5OH
Graphite	Carbon	C
Green vitriol	Ferrous sulfate	$FeSO_4$
Gypsum	Calcium sulfate	$CaSO_4.\ 2H_2O$
	H	
Horn silver	Silver chloride	$AgCl$
Household ammonia	Ammonium hydroxide	NH_4OH
Hypo	Sodium thiosulfate	$Na_2S_2O_3.\ 5H_2O$
	K	
Kaolin	Hydrogen aluminum silicate	$H_2Al_2(SiO_4)_2.H_2O$
	L	
Lampblack	Carbon (impure)	C
Lanolin	Cholesterol	$C_{27}H_{46}O$
Laughing gas	Nitrous oxide	N_2O
Levulose	Fructose	$C_6H_{12}O_6$
Lime, hydrated	Calcium hydroxide	$Ca(OH)_2$
Lime, quick	Calcium oxide	CaO
Lime, slaked	Calcium hydroxide	$Ca(OH)_2$
Limestone	Calcium carbonate	$CaCO_3$

GLOSSARY OF COMMON SUBSTANCES (CONT'D)

COMMON NAME	CHEMICAL NAME	FORMULA
Limewater	Calcium hydroxide solution	$Ca(OH)_2$
Litharge	Lead oxide	PbO
Lithopone	Zinc sulfide and barium sulfate	ZnS and $BaSO_4$
Lunar caustic	Silver nitrate	$AgNO_3$
Lye	Sodium hydroxide	$NaOH$
	M	
Magnesia	Magnesium oxide	MgO
Magnesite	Magnesium carbonate	$MgCO_3$
Malachite	Basic copper carbonate	$CuCo_3.Cu(OH)_2$
Marble	Calcium carbonate (impure)	$CaCO_3$
Marsh gas	Methane	CH_4
Methanol	Methyl alcohol	CH_3OH
Minium	Lead tetroxide	Pb_3O_4
Moth balls	Naphthalene	$C_{10}H_8$
Muriate of potash	Potassium chloride	KCl
Muriatic acid	Hydrochloric acid	HCl
	N	
Niter	Potassium nitrate	KNO_3
	O	
Oil of bitter almonds	Benzaldehyde	C_6H_5CHO
Oil of vitriol	Sulfuric acid	H_2SO_4
Oil of winter-green	Methyl salicylate	$CH_3COOC_6H_4OH$
Oileum	Fuming sulfuric acid	$H_2SO_4.SO_3$
	P	
Pearl	Calcium carbonate	$CaCO_3$
Phosgene	Carbonyl chloride	$COCl_2$
Plaster of Paris	Calcium sulfate	$2CaSO_4.\ H_2O$
Potash	Potassium carbonate or potassium hydroxide	K_2CO_3 or KOH
Pyrolusite	Manganese dioxide	MnO_2
	Q	
Quicklime	Calcium oxide	CaO
Quicksilver	Mercury	Hg
	R	
Rochelle salt	Potassium sodium tartrate	$KNaC_4H_4O_6$
Route	Ferric oxide	Fe_2O_3
	S	
Sal ammoniac	Ammonium chloride	NH_4Cl
Sal soda	Sodium carbonate	Na_2CO_3
Salt of sorrell	Potassium acid oxalate	$KHC_2O_4.H_2O$
Saltpeter	Sodium nitrate	$NaNO_3$
Sand	Silicon dioxide	SiO_2
Silica	Silicon dioxide	SiO_2
Slaked lime	Calcium hydroxide	$Ca(OH)_2$
Soda lime	Calcium oxide & sodium hydroxide	CaO and $NaOH$
Soap (one kind)	Sodium stearate	$C_{17}H_{35}COONa$
Spirit of hartshorn	Ammonium hydroxide	NH_4OH
Spirit of wine	Ethyl alcohol	C_2H_5OH
Sugar	Sucrose	$C_{12}H_{22}O_{11}$
Sugar of lead	Lead acetate	$Pb(C_2H_3O_2)$
Superphosphate	Calcium sulfate & calcium acid phosphate	$CaSO_4$ and $Ca(H_2PO_4)_2$

GLOSSARY OF COMMON SUBSTANCES (CONT'D)

COMMON NAME	CHEMICAL NAME	FORMULA
	T	
TABLE SALT	Sodium chloride	NaCl
TALC	Hydrated magnesium silicate	$3\ MgO\ .4SiO_2.H_2O$
TARTAR EMETIC	Potassium antimonyl tartrate	$2\ K(SbO)C_4H_4O_6.H_2O$
	V	
Vinegar	Acetic acid, dilute	$HC_2H_3O_2$
	W	
Water glass	Sodium silicate	Na_2SiO_3
White lead	Basic lead carbonate	$2PbCO_3.Pb(OH)_2$
White vitriol	Zinc sulfate	$ZnSO_4$
Whiting	Calcium carbonate	$CaCO_3$
Wood alcohol	Methyl alcohol	CH_3OH
	Z	
Zinc blende	Zinc sulfide	ZnS
Zinc white	Zinc oxide	ZnO

CHEMICAL ELEMENTS AND SYMBOLS

NAME OF ELEMENT	SYMBOL	ATOMIC NUMBER	ATOMIC WEIGHT
		A	
Actinium	AC	89	227.0
Aluminum	Al	13	26.98
Americium	Am	95	[243]
Antimony	Sb	51	121.76
Argon	A	18	39.944
Arsenic	As	33	74.91
Astatine	At	85	[210]
		B	
Barium	Ba	56	137.36
Berkelium	Bk	97	[245]
Beryllium	Be	4	9.013
Bismuth	Bi	83	209.00
Boron	B	5	10.82
Bromine	Br	35	79.916
		C	
Cadmium	Cd	48	112.41
Calcium	Ca	20	40.08
Californium	Cf	98	[246]
Carbon	C	6	12.011
Cerium	Cc	58	140.13
Cesium	Cs	55	132.91
Chlorine	Cl	17	35.457
Chromium	Cr	24	52.01
Cobalt	Co	27	58.94
Copper	Cu	29	63.54
Curium	Cm	96	[243]
		D	
Dysprosium	Dy	66	162.46
		E	
Erbium	Er	68	167.2
Europium	Eu	63	152.0

CHEMICAL ELEMENTS AND SYMBOLS (CONT'D)

NAME OF ELEMENT	SYMBOL	ATOMIC NUMBER	ATOMIC WEIGHT
		F	
Fluorine	F	9	19.00
Francium	Fr	87	[223]
		G	
Gadolinium	Gd	64	156.9
Gallium	Ga	31	69.72
Germanium	Ge	32	72.60
Gold	Au	79	197.0
		H	
Hafnium	Hf	72	178.6
Helium	He	2	4.003
Holmium	Ho	67	164.94
Hydrogen	H	1	1.0080
		I	
Indium	In	49	114.76
Iodine	I	53	126.91
Iridium	Ir	77	192.2
Iron	Fe	26	55.85
		K	
Krypton	Kr	36	83.80
		L	
Lanthanum	La	57	138.92
Lead	Pb	82	207.21
Lithium	Li	3	6.940
Lutetium	Lu	71	174.99
		M	
Magnesium	Mg	12	24.32
Manganese	Mn	25	54.94
Mercury	Hg	80	200.61
Molybdenum	Mo	42	95.95
		N	
Neodymium	Nd	60	144.27
Neon	Ne	10	20.183
Neptunium	Np	93	[237]
Nickel	Ni	28	58.69
Niobium	Nb	41	92.91
Nitrogen	N	7	14.008
		O	
Osmium	Os	76	190.2
Oxygen	O	8	16.0000
		P	
Palladium	Pd	46	106.7
Phosphorus	P	15	30.975
Platinum	Pt	78	195.23
Plutonium	Pu	94	[242]
Polonium	Po	84	210.0
Potassium	K	19	39.100
Praseodymium	Pr	59	140.92
Promethium	Pm	61	[145]
Protactinium	Pa	91	231.

CHEMICAL ELEMENTS AND SYMBOLS (CONT'D)

NAME OF ELEMENT	SYMBOL	ATOMIC NUMBER	ATOMIC WEIGHT
		R	
Radium	Ra	88	226.05
Radon	Rn	86	222.
Rhenium	Re	75	186.31
Rhodium	Rh	45	102.91
Rubidium	Rb	37	85.48
Ruthenium	Ru	44	101.1
		S	
Samarium	Sm	62	150.43
Scandium	Sc	21	44.96
Selenium	Se	34	78.96
Silicon	Si	14	28.09
Silver	Ag	47	107.880
Sodium	Na	11	22.991
Strontium	Sr	38	87.63
Sulfur	S	16	32.066
		T	
Tantalum	Ta	73	180.95
Technetium	Tc	43	[99]
Tellurium	Te	52	127.61
Terbium	Tb	65	158.93
Thallium	Tl	81	204.39
Thorium	Th	90	232.05
Thulium	Tm	69	168.94
Tin	Sn	50	118.70
Titanium	Ti	22	47.90
Tungsten	W	74	183.92
		U	
Uranium	U	92	238.07
		V	
Vanadium	V	23	50.95
		X	
Xenon	Xe	54	131.3
		Y	
Ytterbium	Yb	70	173.04
Yttrium	Y	39	88.92
		Z	
Zinc	Zn	30	65.38
Zirconium	Zr	40	91.22

(NOTE: Brackets indicate the isotope of longest known half-life.)

PERIODIC TABLE OF THE ELEMENTS

IA	IIA	IIIB	IVB	VB	VIB	VIIB	VIIIB			IB	IIB	IIIA	IVA	VA	VIA	VIIA	Noble gases
1 H 1.008																	2 He 4.003
3 Li 6.941	4 Be 9.012											5 B 10.81	6 C 12.011	7 N 14.007	8 O 15.999	9 F 18.998	10 Ne 20.179
11 Na 22.990	12 Mg 24.305											13 Al 26.982	14 Si 28.086	15 P 30.974	16 S 32.06	17 Cl 35.453	18 Ar 39.948
19 K 39.102	20 Ca 40.08	21 Sc 44.956	22 Ti 47.90	23 V 50.941	24 Cr 51.996	25 Mn 54.938	26 Fe 55.847	27 Co 58.933	28 Ni 58.71	29 Cu 63.546	30 Zn 65.37	31 Ga 69.72	32 Ge 72.59	33 As 74.922	34 Se 78.96	35 Br 79.904	36 Kr 83.80
37 Rb 85.468	38 Sr 87.62	39 Y 88.906	40 Zr 91.22	41 Nb 92.906	42 Mo 95.94	43 Tc 98.602	44 Ru 101.07	45 Rh 102.905	46 Pd 106.4	47 Ag 107.868	48 Cd 112.40	49 In 114.82	50 Sn 118.69	51 Sb 121.75	52 Te 127.60	53 I 126.905	54 Xe 131.30
55 Cs 132.905	56 Ba 137.34	57 La 138.905	72 Hf 178.49	73 Ta 180.948	74 W 183.85	75 Re 186.2	76 Os 190.2	77 Ir 192.22	78 Pt 195.09	79 Au 196.966	80 Hg 200.59	81 Tl 204.37	82 Pb 207.19	83 Bi 208.2	84 Po (∼210)	85 At ∼210	86 Rn (∼222)
87 Fr (223)	88 Ra 226.02	89 Ac (227)	104	105													

58 Ce 140.12	59 Pr 140.907	60 Nd 144.24	61 Pm (145)	62 Sm 150.4	63 Eu 151.96	64 Gd 157.25	65 Tb 158.925	66 Dy 162.50	67 Ho 164.930	68 Er 167.26	69 Tm 168.934	70 Yb 173.04	71 Lu 174.97
90 Th 232.038	91 Pa 231.036	92 U 238.029	93 Np 237.048	94 Pu (244)	95 Am (243)	96 Cm (247)	97 Bk (247)	98 Cf (251)	99 Es (254)	100 Fm (257)	101 Md (256)	102 No (254)	103 Lr (257)

ANSWER SHEET

TEST NO. ________ PART ______ TITLE OF POSITION ____________________
(AS GIVEN IN EXAMINATION ANNOUNCEMENT - INCLUDE OPTION, IF ANY)

PLACE OF EXAMINATION ____________________ (CITY OR TOWN) ____________________ (STATE) DATE ________

RATING

USE THE SPECIAL PENCIL. MAKE GLOSSY BLACK MARKS.

1	A	B	C	D	E	26	A	B	C	D	E	51	A	B	C	D	E	76	A	B	C	D	E	101	A	B	C	D	E
2	A	B	C	D	E	27	A	B	C	D	E	52	A	B	C	D	E	77	A	B	C	D	E	102	A	B	C	D	E
3	A	B	C	D	E	28	A	B	C	D	E	53	A	B	C	D	E	78	A	B	C	D	E	103	A	B	C	D	E
4	A	B	C	D	E	29	A	B	C	D	E	54	A	B	C	D	E	79	A	B	C	D	E	104	A	B	C	D	E
5	A	B	C	D	E	30	A	B	C	D	E	55	A	B	C	D	E	80	A	B	C	D	E	105	A	B	C	D	E
6	A	B	C	D	E	31	A	B	C	D	E	56	A	B	C	D	E	81	A	B	C	D	E	106	A	B	C	D	E
7	A	B	C	D	E	32	A	B	C	D	E	57	A	B	C	D	E	82	A	B	C	D	E	107	A	B	C	D	E
8	A	B	C	D	E	33	A	B	C	D	E	58	A	B	C	D	E	83	A	B	C	D	E	108	A	B	C	D	E
9	A	B	C	D	E	34	A	B	C	D	E	59	A	B	C	D	E	84	A	B	C	D	E	109	A	B	C	D	E
10	A	B	C	D	E	35	A	B	C	D	E	60	A	B	C	D	E	85	A	B	C	D	E	110	A	B	C	D	E

Make only ONE mark for each answer. Additional and stray marks may be counted as mistakes. In making corrections, erase errors COMPLETELY.

11	A	B	C	D	E	36	A	B	C	D	E	61	A	B	C	D	E	86	A	B	C	D	E	111	A	B	C	D	E
12	A	B	C	D	E	37	A	B	C	D	E	62	A	B	C	D	E	87	A	B	C	D	E	112	A	B	C	D	E
13	A	B	C	D	E	38	A	B	C	D	E	63	A	B	C	D	E	88	A	B	C	D	E	113	A	B	C	D	E
14	A	B	C	D	E	39	A	B	C	D	E	64	A	B	C	D	E	89	A	B	C	D	E	114	A	B	C	D	E
15	A	B	C	D	E	40	A	B	C	D	E	65	A	B	C	D	E	90	A	B	C	D	E	115	A	B	C	D	E
16	A	B	C	D	E	41	A	B	C	D	E	66	A	B	C	D	E	91	A	B	C	D	E	116	A	B	C	D	E
17	A	B	C	D	E	42	A	B	C	D	E	67	A	B	C	D	E	92	A	B	C	D	E	117	A	B	C	D	E
18	A	B	C	D	E	43	A	B	C	D	E	68	A	B	C	D	E	93	A	B	C	D	E	118	A	B	C	D	E
19	A	B	C	D	E	44	A	B	C	D	E	69	A	B	C	D	E	94	A	B	C	D	E	119	A	B	C	D	E
20	A	B	C	D	E	45	A	B	C	D	E	70	A	B	C	D	E	95	A	B	C	D	E	120	A	B	C	D	E
21	A	B	C	D	E	46	A	B	C	D	E	71	A	B	C	D	E	96	A	B	C	D	E	121	A	B	C	D	E
22	A	B	C	D	E	47	A	B	C	D	E	72	A	B	C	D	E	97	A	B	C	D	E	122	A	B	C	D	E
23	A	B	C	D	E	48	A	B	C	D	E	73	A	B	C	D	E	98	A	B	C	D	E	123	A	B	C	D	E
24	A	B	C	D	E	49	A	B	C	D	E	74	A	B	C	D	E	99	A	B	C	D	E	124	A	B	C	D	E
25	A	B	C	D	E	50	A	B	C	D	E	75	A	B	C	D	E	100	A	B	C	D	E	125	A	B	C	D	E

ANSWER SHEET

TEST NO. ____________ PART ________ TITLE OF POSITION ______________________________
(AS GIVEN IN EXAMINATION ANNOUNCEMENT - INCLUDE OPTION, IF ANY)

PLACE OF EXAMINATION ______________________________ (CITY OR TOWN) ______________ (STATE) DATE ____________

RATING

USE THE SPECIAL PENCIL. MAKE GLOSSY BLACK MARKS.

1 A B C D E	26 A B C D E	51 A B C D E	76 A B C D E	101 A B C D E
2 A B C D E	27 A B C D E	52 A B C D E	77 A B C D E	102 A B C D E
3 A B C D E	28 A B C D E	53 A B C D E	78 A B C D E	103 A B C D E
4 A B C D E	29 A B C D E	54 A B C D E	79 A B C D E	104 A B C D E
5 A B C D E	30 A B C D E	55 A B C D E	80 A B C D E	105 A B C D E
6 A B C D E	31 A B C D E	56 A B C D E	81 A B C D E	106 A B C D E
7 A B C D E	32 A B C D E	57 A B C D E	82 A B C D E	107 A B C D E
8 A B C D E	33 A B C D E	58 A B C D E	83 A B C D E	108 A B C D E
9 A B C D E	34 A B C D E	59 A B C D E	84 A B C D E	109 A B C D E
10 A B C D E	35 A B C D E	60 A B C D E	85 A B C D E	110 A B C D E

Make only ONE mark for each answer. Additional and stray marks may be counted as mistakes. In making corrections, erase errors COMPLETELY.

11 A B C D E	36 A B C D E	61 A B C D E	86 A B C D E	111 A B C D E
12 A B C D E	37 A B C D E	62 A B C D E	87 A B C D E	112 A B C D E
13 A B C D E	38 A B C D E	63 A B C D E	88 A B C D E	113 A B C D E
14 A B C D E	39 A B C D E	64 A B C D E	89 A B C D E	114 A B C D E
15 A B C D E	40 A B C D E	65 A B C D E	90 A B C D E	115 A B C D E
16 A B C D E	41 A B C D E	66 A B C D E	91 A B C D E	116 A B C D E
17 A B C D E	42 A B C D E	67 A B C D E	92 A B C D E	117 A B C D E
18 A B C D E	43 A B C D E	68 A B C D E	93 A B C D E	118 A B C D E
19 A B C D E	44 A B C D E	69 A B C D E	94 A B C D E	119 A B C D E
20 A B C D E	45 A B C D E	70 A B C D E	95 A B C D E	120 A B C D E
21 A B C D E	46 A B C D E	71 A B C D E	96 A B C D E	121 A B C D E
22 A B C D E	47 A B C D E	72 A B C D E	97 A B C D E	122 A B C D E
23 A B C D E	48 A B C D E	73 A B C D E	98 A B C D E	123 A B C D E
24 A B C D E	49 A B C D E	74 A B C D E	99 A B C D E	124 A B C D E
25 A B C D E	50 A B C D E	75 A B C D E	100 A B C D E	125 A B C D E